TOPICS IN HYDROMETEROLOGY

Edited by **Theodore V Hromadka II**
and **Prasada Rao**

Topics in Hydrometerology
http://dx.doi.org/10.5772/intechopen.75854
Edited by Theodore V Hromadka II and Prasada Rao

Contributors

Megersa Olumana Dinka, Abderrazak Bannari, Hassan Rhinane, Hicham Bahi, Guilherme Henrique Cavazzana, Fernando J C Magalhães Filho, Ana Paula S Teles, Diego A Zanoni, Susana P Moreira, Denilson O Guilherme, Denise Taffarello, Eduardo Mendiondo, Maria Do Carmo Calijuri, Davi Cunha, Diego Alejandro Guzman Arias, Krishna Gopal Ghosh, Theodore V Hromadka II, Prasada Rao

First published in London, United Kingdom, 2019 by IntechOpen
IntechOpen is the global imprint of INTECHOPEN LIMITED, registered in England and Wales, registration number: 11086078, The Shard, 25th floor, 32 London Bridge Street
London, SE19SG – United Kingdom
Printed in Croatia

British Library Cataloguing-in-Publication Data
A catalogue record for this book is available from the British Library

Additional hard and PDF copies can be obtained from orders@intechopen.com

Topics in Hydrometerology, Edited by Theodore V Hromadka II and Prasada Rao
p. cm.
Print ISBN 978-1-83880-560-9
Online ISBN 978-1-83880-561-6
eBook (PDF) ISBN 978-1-83880-771-9

For EU product safety concerns:
IN TECH d.o.o., Prolaz Marije Krucifikse Kozulić 3, 51000 Rijeka, Croatia,
info@intechopen.com or visit our website at intechopen.com.

Meet the editors

Hromadka & Associates' principal and founder, Theodore Hromadka II, PhD, PhD, PhD, PH, PE, has extensive scientific, engineering, expert witness, and litigation support experience. His frequently referenced scientific contributions to the hydrologic, earth, and atmospheric sciences have been widely published in peer-reviewed scientific literature, including 30 books and more than 500 scientific papers, book chapters, and government reports. His professional engineering experience includes supervision and development of over 1500 engineering studies. He is currently a faculty member at the United States Military Academy at West Point, New York.

Prasada Rao, PhD, is a professor in the Civil and Environmental Engineering Department at California State University, Fullerton. His current research areas relate to climate change, surface and subsurface flow modeling, and computational mathematics. He is also the Associate Director for the International Institute for Computational Engineering Mathematics.

Contents

Preface IX

Chapter 1 **Introductory Chapter: An Introduction to Topics in Hydrometeorology 1**
Theodore V. Hromadka II and Prasada Rao

Chapter 2 **Synergy between SMOS-MIRAS and Landsat-OLI/TIRS Data for Soil Moisture Mapping before, during, and after Flash-Flood Storm in Southwestern Morocco 5**
Abderrazak Bannari, Hassan Rhinane and Hicham Bahi

Chapter 3 **Development and Application of Conceptual Rainfall-Altitude Regression Model: The Case of Matahara Area (Ethiopia) 29**
Megersa Olumana Dinka

Chapter 4 **Linkages between Water and Forests in South American Watersheds under Restoration 39**
Denise Taffarello, Diego Alejandro Guzman Arias, Danielle de Almeida Bressiani, Davi Gasparini Fernandes Cunha, Maria do Carmo Calijuri and Eduardo Mario Mendiondo

Chapter 5 **Geo-Statistical Assessment of the Intensity, Duration, Frequency and Trend of Drought over Gangetic West Bengal, Eastern India 63**
Krishna Gopal Ghosh

Chapter 6 **Rainfall Erosivity: Gap-Filling Method Differences in the Brazilian Pantanal and Cerrado Biomes 83**
Diego A. Zanoni, Susana P. Moreira, Ana Paula S. Teles, Guilherme H. Cavazzana, Denilson O. Guilherme and Fernando JC. Magalhães Filho

Preface

The subject of hydrometeorology encompasses a wide variety of topics that are of high interest in study programs involving the atmospheric sciences, the transport of moisture in the atmosphere, surface water, soil-water hydrology and moisture transport, and earth-surface and atmospheric interactions. Additionally, building on the available computer power, this field has witnessed an exponential increase in applications of computational engineering mathematics (CEM) that has changed the interaction between hydrometeorology and the typical consumer. Weather reports typically include examples of CEM applications such as displays of Doppler radar and illustrations of airflow and other attributes, all developed by modern CEM techniques.

The challenges to overcome for a better understanding of the umbrella of topics that fall under hydrometeorology require multidisciplinary research. Since these topics have a direct impact on human activities, efforts aimed at analyzing the role of weather and climate on the above and related topics can advance the knowledge of hydrometeorology.

Hydrometeorology is a wide-ranging subject and it is difficult for any single book to detail all the associated aspects. In this book, multiple experts present their work on some of these topics. We hope that the topics discussed in this book will motivate researchers to realize the challenges and complexities associated with addressing issues in hydrometeorology and contribute their share toward advancing the knowledge in this upcoming area.

Theodore V. Hromadka II
Professor, Department of Mathematical Sciences
United States Military Academy
West Point, New York, USA

Prasada Rao
Professor, Department of Civil and Environmental Engineering
California State University Fullerton
Fullerton, CA, USA

Chapter 1

Introductory Chapter: An Introduction to Topics in Hydrometeorology

Theodore V. Hromadka II and Prasada Rao

Additional information is available at the end of the chapter

http://dx.doi.org/10.5772/intechopen.85025

1. Introduction

Hydrometeorology is the study of both the atmospheric and terrestrial phases of the hydrological cycle, with emphasis on the interrelationship between them (i.e. the transfers of water and energy between the land surface and the lower atmosphere). Accordingly, the science of hydrometeorology bridges across both hydrology and meteorology [1]. The subject of hydrometeorology encompasses a wide variety of topics [2, 3] that are of high interest in programs involving the atmospheric sciences, the transport of moisture in the atmosphere, surface-water and soil-water hydrology, earth-surface and atmospheric interactions, and techniques in engineering mathematics including the evolving applications in Computational Engineering Mathematics or "CEM". Recent advances in CEM are opening new opportunities for researchers to address diverse challenging multi-dimensional applications in hydrometeorology. Because of the massive database sizes involved, both in hydrometeorology data collection (greatly augmented by private observers equipped with highly accurate monitoring reporting equipment connected to the web central databases), the evolving field of "visualization" has developed where focus is upon generating depictions of these data to increase understanding [4], and to assemble the data for further assessment and analysis for subsequent detailing and publication.

In the current book, a selection of data interpretations and experiences are presented as an enticement to students and practitioners for motivating further intellectual growth in the growing field of study known as "hydrometeorology." Many universities borrow specific topics from the field of hydrometeorology in related courses such as Computational Engineering Mathematics or its earlier version, Engineering Mathematics (or other variants). Perhaps few other demonstrations of vector calculus are as noteworthy and understandable as the description of the vector calculus topics of vector curl and vector divergence, as by examining the evolution of a weather tornado or hurricane, or of the use of these vector concepts in describing weather systems as they move over the planet. The notion of wind flow velocity readily leads to the description of multidimensional vectors, and the readily observable effects of wind rotation as well as expansion and contraction of a moving stream tube of air often invite the student or practitioner or hobbyist, to seek further understanding of the mathematical underpinnings of these vector concepts.

Of particular value are the numerous weather hobbyists [5] as well as the more detailed and validated databases [6, 7] that are increasingly made available for general use. Several publications containing highly descriptive diagrams and remarkable photographs are now available both in hard copy and in the Internet, further exciting the observer and hydro meteorologist to explore the various ways to describe the planet's weather systems and related hydrologic and earth-surface interactions.

The well-known "Moore's Law" truly has application in this area of investigation in that computational advances in power and mathematical capabilities have created an innovation-rich environment where the understanding of the global transport systems can be better interrelated and their interdependence linkages better identified and described. In the following, a very brief introduction is made to each of the chapters in the book. The editors celebrate the assemblage of knowledge achieved by these chapter authors and coauthors.

Bannari et al. used SMOS, OLI, and TRIS data to study the soil moisture that characterizes over the Guelmim city and its neighborhood in the Southwestern of Morocco. This area has very limited rainfall, and when it rains, it is intense over a short period, leading to flash floods. The end results were compared with published data from NOAA climate prediction center. They have described in detail their algorithm which can help other researchers to extend its application for varying geographical domains. The approach used can be integrated with other applications and can help urban planners to design contingencies for any future flooding event.

Although the rainfall measured from gages are reliable, their sparse spatial distribution warrants developing techniques by which the measured gage rainfall values across a network can be extended and corrected to arrive at rainfall estimate at other locations. Correction to the measured values by incorporating altitude and topographical variation can provide reasonable precipitation estimates for planning and design purposes. Dinka addressed this problem by developing a conceptual regression model for the Matahara area, which is located in the middle of Awash valley, about 200 km south east of Addis Ababa, Ethiopia. The relationship between monthly rainfall totals and altitude over Matahara region was examined using the optimized ordinary least square method.

Taffarello et al. presented the hydrometeorological characteristics of few chosen watersheds from Brazil. The challenges in developing baseline performance data across various watersheds that have varying vegetal, anthropic, geological, topographical, land use patterns, biodiversity, and soil characteristics have been discussed. The linkages between ecosystem and soil moisture were reviewed. Salient data from chosen ecosystem-based adaptation (EbA) projects that are being implemented for restoring the watersheds have been presented. Their effort can translate to a valuable tool for decision-makers to review the efficacy of existing best watershed management practices aimed at improving the capacity of aquifers and in devising new practices that can be effectively implemented by integrating all the concerned parties.

Ghosh assessed the spatial and temporal drought intensity during 1991–2002 for the Gangetic region in eastern part of India. The standard precipitation index was used in the analysis together with seven parameters to evaluate drought. The analysis showed that the deficit precipitation in this region since the 1950s has been on upward trend, thus affecting the local socioeconomic fabric in the society. This study sheds more reliable information related to drought that can help area managers to streamline their drought preparedness strategies. Additionally, this approach can be applied to other regions in the world, which rely largely on rainfall for their agricultural needs.

Any effort aimed at conserving soil requires a reliable approach to predict soil erosion, the accuracy of which depends on the input precipitation value. Since at many rainfall monitoring stations, the recorded data are not continuous which need to be first addressed (Zanoni et al.). Three methods (weighted likelihood, multiple regression, and weighted likelihood based on multiple regression) for filling the gaps in measured rainfall values in Brazil watersheds were analyzed. Filling in the rainfall gaps translated to a continuous data for 2001–2004 period from which the rainfall erosivity was calculated at chosen stations in Parana river basin. The erosion values from the three methods were compared, and weighted likelihood method was recommended for further analysis.

Author details

Theodore V. Hromadka II[1]* and Prasada Rao[2]

*Address all correspondence to: ted@phdphdphd.com

1 Department of Mathematical Sciences, United States Military Academy, West Point, New York, USA

2 Department of Civil and Environmental Engineering, California State University, Fullerton, CA, USA

References

[1] University of Arizona, Meteorology, Hydrology, and Hydrometeorology, https://has.arizona.edu/meteorology-hydrology-and-hydrometeorology (Retrieved 7 April 2019)

[2] Bruce JP, Clark RH. Introduction to Hydrometeorology. Oxford: Pergamon Press; 1966

[3] Collier CG. Hydrometeorology. John Wiley & Sons; 2016

[4] Rautenhaus M, et al. Visualization in meteorology - A survey of techniques and tools for data analysis tasks. IEEE Transactions on Visualization and Computer Graphics. 2017;**24**(12):3268-3296

[5] Wunder Blog: Weather Underground. http://www.wunderground.com (Retrieved 7 April 2019)

[6] World Meteorological Organization. http://worldweather.wmo.int/en/home.html (Retrieved 2 April 2019)

[7] National Centers for Environmental Information, NOAA. https://www.ncdc.noaa.gov/cdo-web/ (Retrieved 9 April 2019)

Chapter 2

Synergy between SMOS-MIRAS and Landsat-OLI/TIRS Data for Soil Moisture Mapping before, during, and after Flash-Flood Storm in Southwestern Morocco

Abderrazak Bannari, Hassan Rhinane and Hicham Bahi

Additional information is available at the end of the chapter

http://dx.doi.org/10.5772/intechopen.79554

Abstract

This chapter deals with soil moisture (SM) characterization over the Guelmim city and its neighborhood in the Southwestern Morocco that has been flooded several times over the past 50 years. To achieve this, space-borne SMOS and Landsat-8 OLI/TIRS data were preprocessed to correct several radiometric anomalies, and they were used. The SMOS brightness temperature data acquired before, during, and after the storm with 1-day temporal resolution and coarse spatial resolution (25 km) were transformed to the SM maps. OLI and TIRS data with moderate spatial and temporal resolutions were converted to Normalized Difference Vegetation Index (NDVI) and Land Surface Temperature (LST) to retrieve the Soil Moisture Index (SMI) maps. The results obtained were analyzed, intercompared, and validated against the compiled SM values from rainfall database (SM-RFE) delivered by NOAA climate prediction center Rainfall Estimator (RFE) for Africa. SMOS results show how the spatial variation of SM changes extremely at the regional scale before, during, and after the flash flood day-to-day. The SMI results converge toward the same conclusions showing a drastic SM change before and after flash flood highlighting the impact of inundation and the mud accumulation. By reference to the measured SM-RFE datasets, the validation of the derived SM maps exhibits a significant correlation ($R^2 \geq 0.89$). Globally, we observe a good complementarity among the considered data sources and processing methods for SM spatial information extraction, and the potential of their integration for the development of a prediction and monitoring model for flash flooding at the regional and local scales.

Keywords: soil moisture, SMOS, Landsat OLI/TIRS, land surface temperature, flash flood, storm, inundation, SMI

1. Introduction

Soil moisture (SM) affects the spatial variation of the biosphere and the atmosphere interactions through its influence on the balance of latent and sensible heat flux. It has been adopted by the *Global Climate Observing System* (GCOS) as an essential climate change variable and water resource management [1]. Its estimation and monitoring from space at the spatial and the temporal scales are useful information for hydrological, agricultural, and environmental applications, and modeling. It is required for drought and flood forecasting, natural disaster management, and assistance effort coordination during flooding periods. Moreover, it plays a fundamental role in the prediction of erosion and sediment loads in watershed streams according to topography and lithology [2–4]. In this context and in contrast with many conventional methods, remote sensing science and technology significantly contributed in the activities of SM estimation, flood damage assessment, and landscape management [5, 6]. Indeed, SM information can be derived from optical, thermal, passive, and active microwave remote sensing satellite sensors [2]. Optical sensors such as Landsat *Operational Land Imager* (OLI) exploring the *visible-near-infrared* (VNIR) and the *shortwave infrared* (SWIR) spectral domains characterize the SM based on the absorption feature, i.e., increased SM generally results in a decrease in reflectance. Moreover, the thermal infrared remote sensing, such as Landsat *Thermal Infrared Sensor* (TIRS), related the SM to the soil temperatures based on the heat and the thermal conductivity; thus, variation in SM is automatically related to the land surface temperature [6]. The intensity of passive microwave emission from moist soil is related to its dielectric constant; generally, the increase in SM leads to increase in microwave brightness temperature [7]. For active microwave, the magnitude of the retrodiffusion coefficient is related to the SM through the dielectric constants of soil and water [8]. Currently, *Synthetic Aperture Radar* (SAR) represents the best approach for obtaining spatially distributed surface SM at the medium local scale [2]. Nevertheless, developed by the ESA (*European Space Agency*) in collaboration with the *Centre National d'Études Spatiales* (CNES) in France, and the *Centro para el Desarrollo Tecnológico Industrial* (CDTI) in Spain, *Soil Moisture and Oceanic Salinity* (SMOS) was the first space-borne mission transporting onboard the *Microwave Imaging Radiometer using Aperture Synthesis* (MIRAS) instrument dedicated to near-surface (0–5 cm depth) SM mapping on a global scale using L-band. According to many studies, this band is one of the most promising approaches to monitor SM at the global scale with regular and high temporal resolution [7–9].

Furthermore, during the last four decades, the impact of climate change became an undeniable reality, with a broad consensus of the international scientific community on the significance of its impact on the environment and economic and social factors, especially in African countries. Morocco is a North-African country with a dominant (mostly) semiarid to arid climate and presents typical characteristics of Mediterranean landscapes vulnerable to land degradation processes, landslides, and desertification risks [10, 11]. Currently, Morocco is experiencing the longest dry episode of its contemporary history characterized by a reduction of precipitation and a rise in temperatures. In the southern regions of Morocco, the rainfall rarely occurs but with high intensity during a short period of time, which causes flooding problems and accelerated the erosion phenomenon and land degradation [3, 4]. According to Erskine and Saynor

[12], catastrophic floods are defined as events with a flood peak discharge at a rate of at least 10 times greater than the mean annual flood. Heavy rains often induce floods in Morocco, including flash floods, river floods, and mud floods during the rainy season. Indeed, in November 2014, violent storms caused flooding and impressive river floods in a large part of southern Morocco, especially in the city of Guelmim and regions, which are localized at the foot of the Atlas Mountains, with peaks rising to over 4000 m. According to SIGMA [19], this natural catastrophe caused the death of more than 46 persons and a significant damage to the infrastructure; villages were inundated causing thousands of houses to collapse, many oasis and agricultural fields were destructed, and power and telephone networks and several roads and bridges were damaged (**Figure 1**). Total losses were estimated about 0.6 billion US$ [13]. Consequently, the region of Guelmim was declared a "disaster area" by the Moroccan government. This area was not devastated for the first time; it has been flooded several times over the past 50 years, namely in 1968, 1985, 1989, 2002, 2010, and 2014. Unfortunately, in addition to the climate change impact, this situation happened because of the lack of emergency measures and a failure of development policy from the decision-makers. However, in order to improve the management of water regulation structures, to maximize water storage capacity, and to reduce the risks caused by floods, remote sensing (science and technology) has become a fundamental solution for flood monitoring and its impact assessment through SM [14]. The objective of this chapter is the SM mapping at the regional and local scales during the critical time of flash-flood storm exploring combined approaches based on SMOS-MIRAS and Landsat-8 (OLI and TIRS) datasets, respectively, and coarse and medium spatial resolutions.

Figure 1. Impressive flood and inundated village, destruction of road and bridge infrastructure, and driver rescues by helicopter (Guelmim region on 24 November 2014, photos from the web).

2. Material and method

The used methodology is summarized in **Figure 2**. It involves four fundamental steps: (i) data preprocessing, (ii) SM mapping from SMOS-MIRAS data, (iii) *Soil Moisture Index* (SMI) map retrieval from OLI and TIRS data, and (iv) obtained result analysis, interpretation, intercomparison, and validated against the compiled SM values from rainfall database (SM-RFE) delivered by NOAA climate prediction center *Rainfall Estimator* (RFE) for Africa.

2.1. Study site

Guelmim is a city in the south of Morocco (**Figure 3**) and is located at the foot of the western Anti-Atlas Mountains with peaks rising to over 2400 m above sea level (28° 59′ 02″ N, 10° 03′ 37″ W). It follows the course of underground shallow aquifers and dry rivers and it is characterized by a semiarid and arid subtropical climate. The temperature range varies from 12°C in January to 49°C in July. Annual rainfall averages between 70 and 120 mm/year. The geological formations that feed alluvium are granite, schist, quartzite, sandstone, limestone, dolomite, marl, conglomerate, andesite, and rhyolite. From a geological point of view, this region constitutes a complex synclinal, framed and surrounded in the N, W, and S by three Precambrian

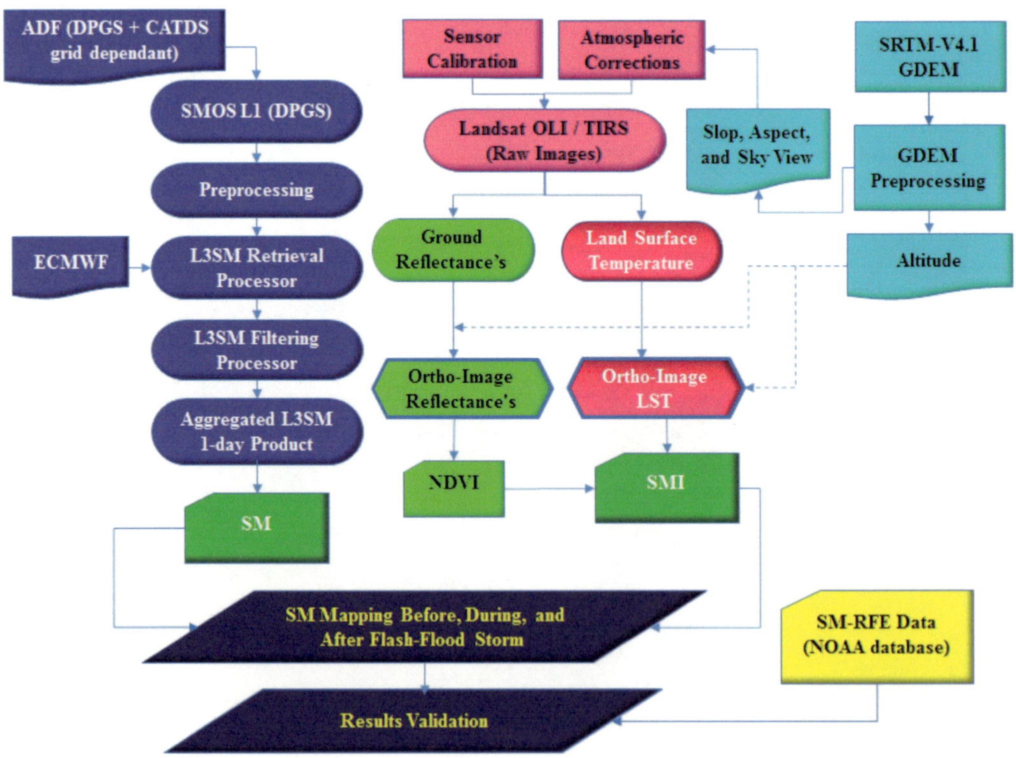

Figure 2. Methodology flowchart.

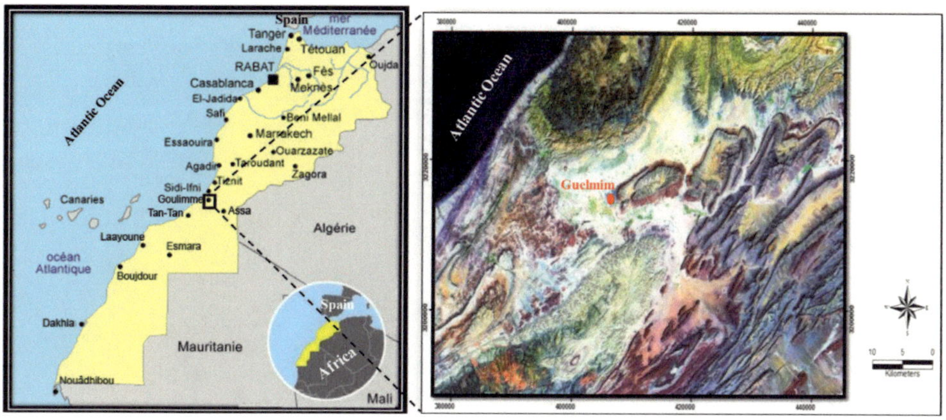

Figure 3. Study site.

anticlinal inlets [15]. The two main structural units in the region are the carbonate plateaus and the folded Bani Hills. The most important Infra-Cambrian and Cambrian carbonate plateaus are located in the north, consisting of a continuous area bordering from W to E the Ifni Inlet, Akhsass plateau, and the southern flank of the Kerdous inlet. The second one, located south, is formed by the external part of Jabal Guir-Taissa. These plateaus are surrounded by schist and sandstone formations of the Georgian age. At their foot begin large and elongated plains consisting of Acadian schist covered by Quaternary deposits. At the center of the Guelmim basin, Jabal Tayert is found, which is formed by green Upper Acadian schist and covered at the top by hard sandstone and quartzite bars. The Bani Jabal is a folded structure consisting of several aligned and NE-SW–oriented synclinals alternating with narrow anticlinals formed by Acadian or Ordovician sandstones and quartzites.

Furthermore, the Guelmim watershed covers a total area of approximately 7000 km^2, forming a network of wadis (rivers) along with several spreading floodwater areas (**Figure 4**). The hydrographic network is made up of three subwatersheds of the following main wadis: wadi Seyyad, wadi Noun, and wadi Oum Al-Achar. Wadi Seyyad originates at an altitude of 1200 m on the southern slopes of the Anti-Atlas Mountain. It flows in an E-W direction, composed of impermeable rocks and mainly receives numerous tributaries of its right bank; its watershed covers about an area of 2860 km^2. Wadi Noun drains the southern area, where the bit is marked with river beds that promote natural flooding. With a length of 143 km, its watershed comprises an area of about 2240 km^2. The wadi Assaka begins in the Akhsass massive at an altitude of 1150 m. It flows through the corridor between the Jebel Adrar and Guelmim west, discharging into the Atlantic Ocean. Finally, Wadi Oum Al-Achar, with a watershed of 1170 km^2, crosses a wide plain of 7 km and is located between the Tayert Hill and Ifni boutonnière. It drains the southern slopes of the Akhsass region, and its main tributaries are located in the plain. All three wadis lie on schistous impermeable large valleys, covered by low permeable Quaternary carbonates and fluviolacustrine silts. The confluence of the three wadis, downstream from Guelmim city, forms Wadi Assaka, which begins in the Akhsass massif at an altitude of 1150 m. It goes through the corridor between the Jabal Adrar and Guelmim west,

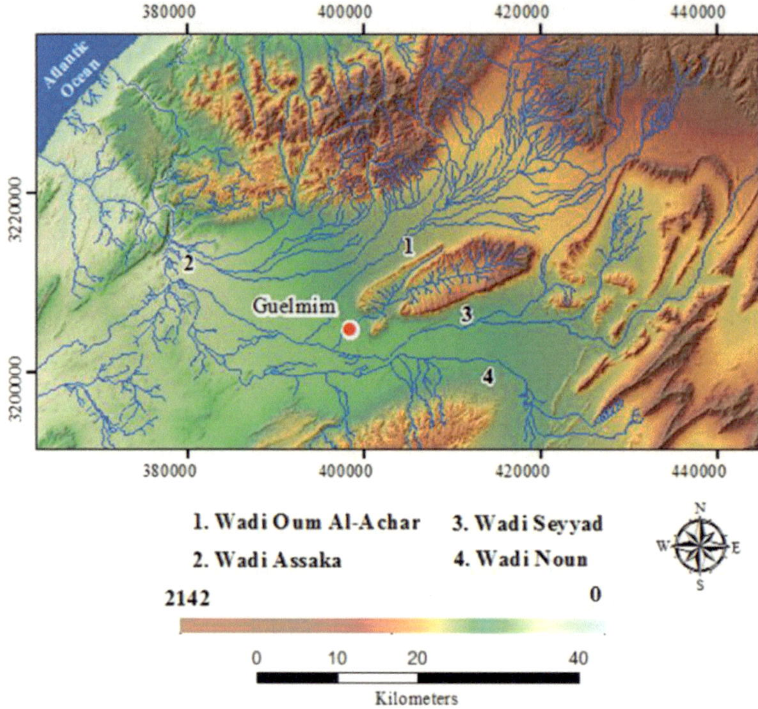

Figure 4. SRTM-V4.1 DEM with 30-m pixel size and the hydrological network overlaid on the study site.

eventually discharging into the Atlantic Ocean after crossing narrow gorges. This hydrographic system is often inactive, especially during the summer, when the flow is very low; however, it becomes active during the winter period (December to March).

2.2. SMOS data

SMOS is the first active remote sensing space-borne mission measuring the near-surface SM and sea surface salinity at the global scale [16]. Successfully launched on November 2, 2009, it is a small satellite-platform transporting onboard a 2D interferometer instrument named MIRAS composed of 69 small L-band receivers (21 cm and 1.4 GHz) measuring the phase difference of radiation in vertical and horizontal (V and H) polarizations under various incident angles with a high temporal resolution [16]. SMOS has a sun-synchronous orbit at 757 km altitude with a 06:00 Local Standard Time $\pm$ 15 min ascending equator crossing time. It provides global SM information with an accuracy better than 0.04 m^3/m^3 [17]. The pixel size of the acquired images is nominally 35 km at the scene center and resampled to 25 km. This information is used in predictive atmospheric, oceanographic, and hydrologic models to understand the climate system and the water cycle, to forecast the extreme natural events (floods and droughts), to manage the water resources adequately, and to improve the climate change impact assessment at the global and regional scales. In this study, the SMOS daily Level 1 (L1) primary brightness temperature (BT) data acquired between November 1 and December 10, 2014, were preprocessed, processed, and downloaded using *Centre Aval de Traitement des Données SMOS*, http://www.catds.fr/Products/Available-products-from-CPDC (CATDS).

2.3. Landsat OLI/TIRS data

Since 1972, the Landsat scientific collaboration program between the NASA and USGS constitutes the continuous record of the Earth's surface reflectivity from space. Indeed, the Landsat satellite series support more than four decades of a global moderate resolution data collection, distribution, and archive of the Earth's continental surfaces to support research, applications, and climate change impact analysis at the global, the regional, and the local scales [18]. On February 11, 2013, the polar-orbiting Landsat-8 satellite was launched, transporting two push-broom instruments: OLI and TIRS. The OLI sensor collects land surface reflectivity in the VNIR and SWIR wavelength regions as well as a panchromatic band. The band passes are narrower in order to minimize atmospheric absorption features [19], especially the NIR spectral band (0.825 μm). Its design results in a more sensitive instrument with a significant amelioration of the *signal-to-noise ratio* (SNR) radiometric performance quantized over a 12-bit dynamic range (Level 1 data); raw data are delivered in 16 bit. This SNR performance and improved radiometric resolution provide a superior dynamic range and reduce saturation problems associated with globally maximizing the range of land surface spectral radiance and, consequently, enable better characterization of land-cover conditions. Furthermore, TIRS sensor incorporates two narrow thermal spectral bands (band 10:10.8 μm and band 11:12 μm) with a spatial resolution of 100 m, but resampled and delivered with 30 m to match OLI bands. These two bands use Quantum Well Infrared Photodetectors (QWIPs) to detect long wavelengths of light emitted by the Earth whose intensity depends on surface temperature. Moreover, they were designed to allow the use of "split-window" surface temperature retrieval algorithms, which allow the distinction between the temperature of the Earth's surface and that of the atmosphere [18]. Because of limited clear-sky conditions and cloud presence over the study site during the flash-flood storm that prevented the optical sensor observations, and the 16 days temporal resolution, only two pairs of raw images acquired by OLI and TIRS sensors were used in this research. The first pair (optic and thermic) was acquired on November 7, 2014, 2 weeks before the flash flood, and the second pair was collected on December 9, 2014, 8 days after the flash flood.

2.4. SRTM-V4.1 DEM data

The *Shuttle Radar Topography Mission* (SRTM) collected the most complete high-resolution digital topographic database over 80% of the Earth's land surface from 60° N to 56° S during an 11-day mission, which was flown aboard the space shuttle *Endeavor* from February 11–22, 2000 [20]. The fundamental objectives of this mission are to provide important information for NASA's Earth Sciences Enterprise, which is dedicated to understanding the total Earth system and the effects of human activity on the global environment [21]. Since 2000, the SRTM data have been provided in 30-m pixel size only within USA territory, while for the rest of the world, the data were available for public use at 90-m pixel size. In January 2015, the US government released globally the full resolution of the original measurements (30-m pixel size). The data are projected in a geographic coordinate system using a WGS-84 geodetic reference and EGM-96 (*Earth Gravitational Model* 1996) vertical datum. According to USGS [21], at 90% confidence, the absolute vertical height accuracy is equal or less than ±16 m, there is a relative vertical height accuracy of less than ±10 m, and there is a circular absolute planimetric error of less than ±20 m and a circular relative planimetric error of less than ±15 m [20]. However, with reference to a topographic contours map (1:50,000), Bannari et al. [22] showed that the derived global height surface accuracy is ±3.15 m in Guelmim region. Moreover, they demonstrated that this accuracy is significantly influenced by topography; error is larger (±11.34 m) for high-altitude terrain with strong slopes, while it is smaller (±1.92 m) in the low-to-medium relief areas with indulgent slopes. This SRTM-V4.1 DEM was used for Landsat image preprocessing and for the study site topographic and hydrographic network characterization.

2.5. Soil moisture from rainfall estimator

Thanks to the close connection between rainfall and SM. Definitely, the ground rainfall measurement provides the accurate method to obtain information about daily rainfall, which is automatically used to estimate SM [23]. However, the spatially sparse network of meteorological stations and the temporally incomplete records at several stations across Africa leaves large parts of the continent unobserved. To resolve this problem, satellite-based algorithms have the advantage of providing full spatial coverage over Africa. The combination of the ground rainfall measurements and the daily satellite-based rainfall observations with contrasting estimation approaches are extremely valuable for rainfall monitoring and SM estimation over African territory. The RFE database setup by NOAA [24] is a product based on an algorithm exploiting four different data sources and several processing methods to estimate precipitation appropriately [24] in order to calculate after the standard water requirement satisfaction index to estimate soil moisture (SM-RFE) over Africa [25]. This independent SM-RFE database was used for the validation of the derived SM maps in this study.

2.6. Data preprocessing

2.6.1. SMOS

The SMOS data products are delivered in four levels. The L0 is the raw measured data with MIRAS radiometer and processed by the *Data Processing Ground Segment* (DPGS). The L1 is the primary BT data converted from the raw data (L0). The levels L2 and L3 products are, respectively, the retrieved SM and vegetation optical depth using an iterative scheme and multiangular BT observations [26]. The fundamental difference between these products (L2 and L3) is that the product L3 considers simultaneously several revisit observations over each pixel (multiorbit retrieval approach) to improve SM estimation [17]. These products are derived based on complex preprocessing chains at CATDS involving *Ancillary Data Files* (ADF), data obtained from SMOS combined with additional data from other sensors or models for calibration (**Figure 2**). In this study, the daily SMOS L3 multiangular BT data products selected from CATDS data center were used to retrieve SM. The data were expressed at the top of the atmosphere (i.e., without correction for select reflected extraterrestrial sky and atmosphere contributions) and at predefined incident angles varying from 2.5 to 62.5° with 5° steps. They were delivered for both ascending and descending orbits, and projected on a global EASE grid with 25 km pixel size. The preprocessing steps consist of the following operations: spatial resolution filtering, *Radio Frequency Interference* (RFI) detection (to exclude data with a probability of RFI larger than 0.2, and urban or water cover fractions larger than 0.1), correction of grid nodes where sun specular effect occurs, computation of radiometric noise correction factor, computation of the initial validation index, and computation of atmospheric and sky contribution [27]. The reader can find more details about the preprocessing steps in Kerr et al. [17].

2.6.2. Landsat OLI/TIRS

Optical sensor calibration and atmospheric corrections are fundamental preprocessing operations to restore the images' radiometric quality. The changes caused by these artifacts can be mistakenly attributed to changes in the land use and ground biophysical components, and errors can propagate in all subsequent image processing steps, such as spectral indices calculations, multitemporal analysis, climate change modeling, etc. [28, 29]. The *Canadian Modified Simulation of a Satellite Signal in the Solar Spectrum* (CAM5S) based on the Herman radiative transfer code [30] was used for atmospheric parameter simulation in OLI spectral bands to calculate all the requested atmospheric correction parameters. To preserve the radiometric integrity of the images, absolute radiometric calibration and atmospheric effects were combined and corrected in one step. Likewise, to eliminate the distortions caused by the relief and the shadow impact, especially in the Guelmim study area with Atlas Mountains peaks of more than 2400 m, an orthorectification was conducted using SRTM-V4.1 GDEM [22]. In addition, topographic attributes such as altitude, slope, aspect, and sky view, integrated into the orthorectification approach, were extracted from this DEM [31]. The Rational-Function Model implemented in the Ortho-Engine module of PCI-Geomatica was used in this step that allows the corrections of the parallax effect at the spatial arrangement of pixels along track, disruptive effects caused by shadow and topographic variability, as well as the residual atmospheric artifacts caused by altitude variability (**Figure 2**). To preserve the image radiometric integrity, geometric corrections have been combined into a single step with the correction of topographic effects [29]. Furthermore, the preprocessing of TIRS data (band 10) was done by conversion of the digital numbers to BT using the Planck radiance function (Eq. (1)) and the thermal calibration constants (K_1 and K_2) provided in the images' metadata files [32]. Then, the BT was transformed to *Land Surface Temperature* (LST), which is a key variable in Earth environment research for calculating the *Soil Moisture Index* (SMI). The LST was calculated based on a method that combines the vegetation cover fraction (P_v) derived from the NDVI and the emissivity (ε), expressed by the Eq. (2) [33]. Based on the NDVI histogram thresholds (Eq. (4)), the vegetation fraction (P_v, Eq. (5)) was estimated from $NDVI_{min}$ and $NDVI_{max}$ referring, respectively, to nonvegetated and very dense vegetated land covers.

$$T_b = \frac{K_2}{\ln\left(\frac{K_1}{L_\lambda} + 1\right)} \tag{1}$$

$$LST = \frac{T_b}{[1 + (\lambda \times T_b/C_2) \times \ln(\varepsilon)]} \tag{2}$$

$$C2 = h \times c/s \tag{3}$$

$$\varepsilon = \begin{cases} \mathrm{a} + \mathrm{b}\,\rho\mathrm{red} & \text{when NDVI} < 0.2 \\ \varepsilon_v \mathrm{P_v} + \varepsilon_s(1 - \mathrm{P_v}) + d\varepsilon & \text{when 0.2 NDVI 0.5} \\ \varepsilon_v + d\varepsilon & \text{when NDVI} > 0.5 \end{cases} \tag{4}$$

$$\mathrm{P_v} = \left[\frac{\mathrm{NDVI} - \mathrm{NDVI_{min}}}{\mathrm{NDVI_{max}} - \mathrm{NDVI_{min}}}\right]^2 \tag{5}$$

$$\mathrm{NDVI} = (\rho_{\mathrm{nir}} - \rho_{\mathrm{red}})/\rho_{\mathrm{nir}} + \rho_{\mathrm{red}}) \tag{6}$$

where T_b is at-satellite BT, λ is the wavelength of emitted radiance, and K_1 (774.89) and K_2 (1321.08) are the calibration constants for the band 10. $C_2 = 1.4388*10^{-2}$ m K and it is presented by Eq. (3). ε is the emissivity computed from Eq. (4) [33]. h is Planck's constant = $6.626*10^{-34}$ J/s, c is the velocity of light = $2.998*10^{8}$ m/s, and s is Boltzmann constant = $1.38 * 10^{-23}$ J/K. ε_v is the vegetation canopy emissivity, ε_s is the bare soil emissivity, and dε is the internal reflection emissivity due to cavity effect, while ρ_{red} and ρ_{nir} are the surface reflectance in the red (OLI-4) and near-infrared (OLI-5) spectral bands, respectively.

2.7. Data processing

2.7.1. SMOS

The multiangular BT data products were processed using the SMOS retrieval algorithm to derive SM. Based on the *L-band Microwave Emission of Biosphere* (L-MEB) radiative transfer model [16], this algorithm considers an iterative approach minimizing the difference between the observed and estimated BT at H and V polarizations, for a variety of incident angles. Moreover, it includes a number of parameterizations to capture effects of vegetation structure and soil roughness on polarization and angular properties of BT emitted from land surfaces [34]. This algorithm has been calibrated and validated several times using *in situ* measurements [26, 35], and recent validation process shows an accuracy around 0.04 m^3/m^3 in comparisons with observations from watershed networks [36]. The soil temperature product derived from the *European Centre for Medium-Range Weather Forecasts* (ECMWF) was introduced in this algorithm to provide a ground reference "calibration-information" useful for the time series analysis to retrieve the most accurate possible SM maps [36]. A multilinear regression approach was achieved over the considered data using calibrating coefficients and several ancillary data extracted from ADF until the algorithm finds the best set of parameters to derive the daily L3 SM maps. More details about the processing process are described in Kerr et al. [17].

2.7.2. OLI and TIRS

SM conditions at different times (i.e., between droughts and flood periods) can change extremely. In the literature, we know that the SM in the first centimeters of the soil surface is strongly related to the emitted BT measured with L-band [37]. Moreover, as mentioned before, other studies have shown the potential of the combination of LST derived from thermal infrared radiation and NDVI to estimate and to monitor SM conditions. The idea behind this combination is the linkage between hydrological cycle (water content in soil) and biophysical parameters (vegetation cover density), and the connection in turn interacts with atmosphere-biosphere coupled system. Exploring this concept, an empirical parametrization was established between LST and NDVI (scatter-plot in 2D space) by Zeng et al. [38] who showed how land surface biophysical properties vary as a function of SM conditions. Then, they proposed the SMI, whose values are ranged from 0 to 1 indicating, respectively, severe drought and very moist conditions. This SMI is defined as follow [38]:

$$SMI = \frac{LST_{max} - LST}{LST_{max} - LST_{min}} \quad (7)$$

$$LST_{max} = a_1.NDVI + b_1 \quad (8)$$

$$LST_{min} = a_2.NDVI + b_2 \quad (9)$$

where LST_{max} and LST_{min} are the maximum and minimum of land surface temperatures for a given NDVI. LST is the observed land surface temperature at the given pixel for a given NDVI. a_i and b_i ($i = 1, 2$) are parameters obtained by the linear regression (a is the slope and b is the intercept) defining both dry and moist edges of the data.

3. Results analysis and discussion

3.1. Topographic and hydrographic network analysis

The topography controls the flow and speed of water, as well as the direction of its dispersion during the inundation. To understand the SM and the flash-flooding mechanisms in the study area, it is so important to analyze the geomorphologic and the topographic variations, and the hydrographic network, and their contribution in these natural hazards. The study site has two main geomorphologic units, the limestone plateau of the Anti-Atlas and quartzite ridges, which is limited by the Atlas Mountains (in the N, S, E, and W) as a natural barrier, which leads to water retention in case of high precipitation intensity. It is characterized by broader valleys and depressions surrounded by hills with heights varying from 153 to 2060 m (**Figure 4**) and steep slopes varying between 9.5 and 26°, which converge toward the interior of the Guelmim plain. The topography of this plain is classified into seven classes whose altitude range vary significantly between 200.0 and 573.5 m, starting from northeast to southwest with approximately 373.5 m height difference. This morphology leads to water retention in the case of rainstorm and, consequently, contributes to the risk of inundation. Thus, it is one of the factors supporting the risk of flash floods. Indeed, as illustrated in **Figure 4**, the hydrographic network is made up of three subwatersheds and four main wadis (wadi Seyyad, wadi Noun, wadi Assaka, and wadi Oum Al-Achar) that flow on a steep slope. It can be seen that the slope orientation and direction of the Guelmim watershed are facing the center of the plain. Moreover, a hill in the East forms a natural barrier with a denivelation of approximately 100 m, which creates a natural basin promoting the accumulation of water and sediments over approximately 14 km distance. The topography variation starting from the foot of Guelmim city (NE-SW) illustrates a very strong slope (26°), which ends on a terrain with concave morphology forming a natural basin. This basin facilitates the accumulation of storm flood, thereby concentrating runoff water, sediment, and mud load. The highest altitude values (between 500 and 2400 m) with the steep slopes and ridges (corresponding to schist and soft Quaternary deposits), representing the streams and drainage system, depressions, and broader-valleys, are related to a relatively high slope gradient ≥20°. They contribute significantly to the erosion's aggressiveness associated with a significant degree of sediment transportation and land degradation risk process. Hard rocks, such as Precambrian quartzite, Adoudounian limestone and dolomite, Ordovician quartzite and sandstone, and Georgian black limestone, characterize these zones. The lowest altitude values represent relatively flat areas with a low slope (≤4°) in the NE-SW direction, which is the hydrographic network direction. In addition, morphological factors influence the susceptibility to flooding and sediment deposition and accumulation, because water tends to flow and accumulate in response to gradients in gravitational potential energy. In general, this first analysis demonstrates clearly that rainfall and topography are the major contributing factors to flash flooding and catastrophic inundation in the study area. The runoff waterpower delivers vulnerable topsoil and contributes strongly to the erosion and land-degradation process after a flood storm in the Guelmim basin. It then transports soil material and sediments to the plain through natural action, i.e., water power and gravity. As illustrated by the photos in **Figure 1** that were acquired during the same day of the flood storm, the water color was dark-red because of its turbidity as it was very rich with sediments and eroded particles. Certainly, the role of the lithology associated with the terrain morphology is decisive in the erosion risk, land degradation, and spatial repartition of SM in this region. This analysis shows how the topographic and hydrographic networks contribute to inundations in the study site and, therefore, highlight the areas subject to different moisture levels.

3.2. SMOS result analysis

As discussed before, using the daily SMOS L3 data, SM maps were generated for each day from the 1st of November to the 10th of December over the Moroccan territory. During these 40 days, **Figure 5** illustrates the SM temporal variation over the Guelmim city and region day-by-day before, during, and after the flood storm. These SM values are the average of four SMOS pixels (2 by 2) covering the study site, which is approximately 50 by 50 km^2. Increased SM values are observed from 2% (1st of November) to 22% (4th of November) during the dry period and after the first precipitations, respectively. Then, the moisture values gradually decreased in the soil from 22 to 4%, respectively, between the 4th and 19th of November. A strong increase of SM took place between the 19th and 21st of November reflecting events with

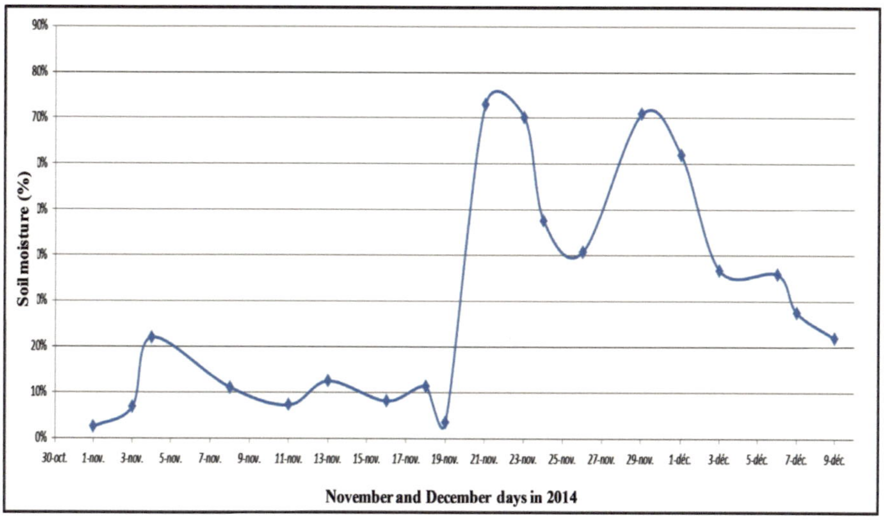

Figure 5. Soil moisture variability over Guelmim region between the 1st of November and the 9th of December 2014 using SMOS satellite data.

a flood peak discharge. It can also be observed that the peaks marked these dates are followed with succession periods of rain maintaining high SM conditions. In fact, during 1 week (19th to 26th of November), SM values remain above 40% with a high peak (SM of 72%) on the 22nd of November. Then, a new flood peak discharge was registered (SM of 74%) on the 29th of November, and progressively the SM values decreased to 23% on the 9th of December. Globally, the highest values (SM > 40%) were recorded between 21st of November and 03rd of December, period when the floods were at their peak. In the following analysis, only six SMOS SM maps reflecting six decisive temporal periods have been chosen to express changes of SM values during these flood events. **Figure 6a** illustrates the SM map for the 2nd of November expressing a severe drought situation over the Moroccan territory, especially over Guelmim region with 2% water content in the soil. On the 07th of November (acquisition date of the first Landsat OLI/TIRS image before the flash flood), the SM value gradually increased to 13% in the southwestern Morocco, but in the center and north regions, the storm had already begun and SM values were, respectively, around 35 and 60%. **Figure 6c** and **d** illustrates the SM maps during the extreme flood events over the majority of the southern Moroccan territory, including Guelmim city and neighborhood; recorded SM values were 73% and 43% for November 21 and 27, 2014, respectively. As a consequence of high precipitation causing very destructible floods, strong changes of SM were also observed on the 1st of December recording 65% water content in the soil (**Figure 6e**). Then, on the 9th of December that coincides with the acquisition date of the second Landsat OLI/TIRS image (8 days after the flash flood), **Figure 6f** shows a decrease in precipitation and the recorded SM content was around 22%. With reference to the calculated SM from SM-RFE NOAA database, the validation of these SMOS SM maps exhibit a significant correlation ($R^2 > 0.90$) at $p < 0.05$ using a first-order polynomial function. This meaning is in agreement with several other results of SMOS validation around the world [36, 39, 40].

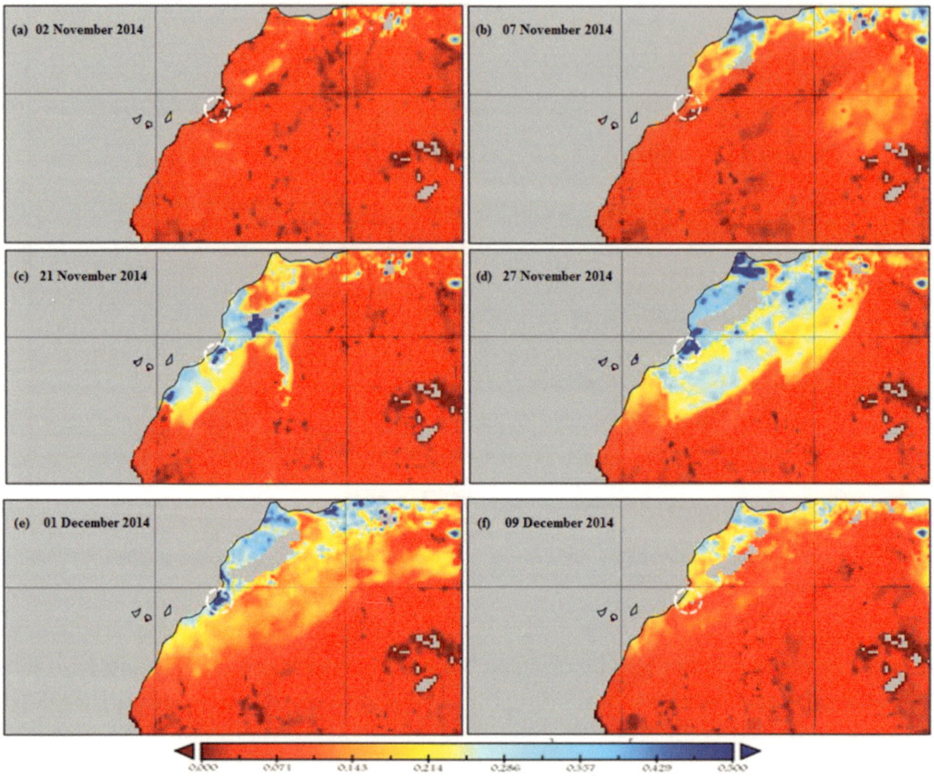

Figure 6. SM maps derived from SMOS data over Morocco (Guelmim region surrounded by the white circle) for November 2, 7, 21, and 27, 2014, as well as for December 1 and 9, 2014.

Furthermore, we observe generally that these derived SMOS maps allow SM characterization and monitoring with excellent temporal resolution over a very large territory or at a country scale, independently to the meteorological conditions. It is possible to predict day-by-day the power, the cycle, and the direction of the storm progress. Obviously, the information is very useful for a warning system development for flash-flooding prediction. Nevertheless, despite these significant and positive potentials, unfortunately, the major disadvantage of SMOS data for this type of application is the coarse spatial resolution that is not sufficient for detailed and accurate information extraction at the locale medium or large scales. For instance, for rescue organization, the roads and river networks are not visible on this kind of data. Moreover, it is impossible to predict the river dynamics and their relation with topographic attributes for relevant risk management in order to organize evacuations of the population and vehicles to the less hazardous spots. Globally, this type of data does not allow us to understand and to study what it is missing in the field (real world) to ensure the safety of population. Moreover, according to high spatial distribution and heterogeneity of SM, there is a need to characterize it at relevant spatial scale to improve the representation of runoff and precipitation in hydrologic and weather prediction models, resulting in improved predictive skills (drought, flood, etc.). Also, the challenge will be mainly to characterize complex mixed pixels including a variety of targets: water, crops, fallow layer, urban, roads, etc. Indeed, the SMOS-MIRAS sensor observes a very large watershed as a whole target in one pixel, and the Guelmim city and regions are covered by 4 pixels that are considering the city as a bare soil. In addition to the pixel complexity, in this mountainous area, SMOS pixels present a mixture of different topographic facets with varying slopes and azimuths, shadowing and adjacency effects, inducing effects which may eventually render the SM estimation less accurate. Indeed, in areas with strong topography such as the Moroccan Atlas Mountains, we know that SM is controlled automatically by gravity-driven water to horizontal regions or areas with very low topographic variation. In such case, SM can be related to topographic features through catchment hydrologic models that require a medium spatial resolution. While, in flat areas, soil mineralogy composition and texture and vegetation cover properties have a stronger impact than topography. Obviously, for addressing the monitoring of SM for flash-flood storm and hydrological applications in areas with high altitude variations and complicated soil mixt-pixels, medium spatial resolution is required. This conclusion is corroborated by several other studies about flood modeling [41], hydrological model calibration [42], and flood simulation [43]. Other studies have revealed that the contribution of satellite remote sensing for SM for runoff modeling is still imprecise [44]; others obtained moderate improvement in hydrological modeling through the assimilation of SM derived from satellite data [45], while others obtained insignificant performances [46]. These distinct results have to be attributed to the inherent uncertainties and issues involved in the use of satellite with coarse spatial resolution for SM estimation and its integration in flood prediction and hydrological modeling. However, other optimistic studies summarize the emerging SM applications, the open issues, and the future opportunities given by new generation of satellites planned for near future [47]. Of course, we share this optimism because, for instance, the synergy between the SMOS coarse spatial resolution and medium spatial resolution of optical and thermal data is likely helpful to achieve a multiresolution SM retrieval approach. Several downscaling methods and integration with medium or low resolutions data such as Landsat-8 or MODIS, as well as using other ancillary data, are required to quantify subpixel heterogeneity of SM. Certainly, the combination and integration in GIS environment of SM subpixel values with the density of the hydrographic and road networks, as well as topographic features (as discussed above), will contribute significantly to inundation monitoring and management.

3.3. MSI result analysis

Figure 7 illustrates the geographic location and density of vegetation cover mapped with NDVI before the flood storm. As a consequence of climate change impact and deforestation in the region, scattered vegetation cover (in green color) is observed in the middle of the Guelmim plateau and along the river beds, represented by small agricultural fields and oases with palm trees. At the top of the mountain (N-W and N-E), a relative high grass and shrub density can be observed, because this high-altitude region benefits from a mild oceanic climate (near to the coast) and the dew, which irrigate significantly this vegetation. Contrariwise, in the south, southeast, and southwest, the vegetal cover is completely absent. Despite the fact that a vegetation index alone is not generally helpful for flood impact assessment or sediment accumulation estimation, the NDVI before and after the flood storm was calculated and integrated with LST for the SMI map derivation. Only the NDVI before flooding is presented

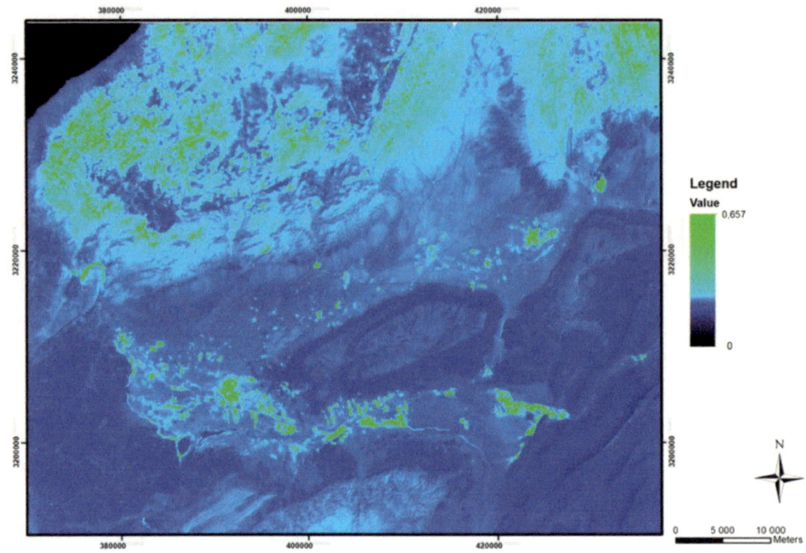

Figure 7. NDVI before the flood storm.

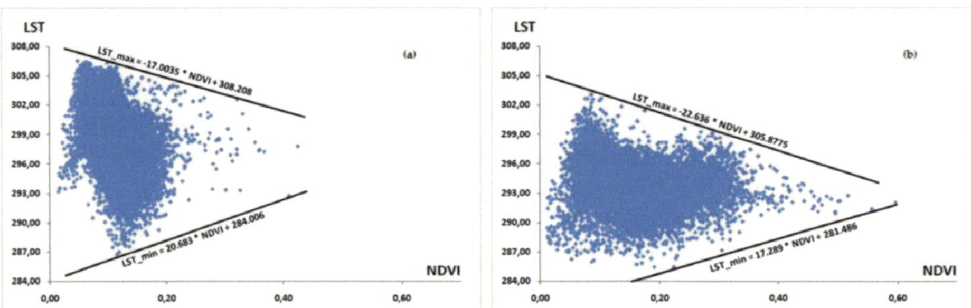

Figure 8. A scatter plot between LST and NDVI before (a) and after (b) the flood storm.

here; after flooding, it showed similar patterns except a very modest increase of herbaceous vegetation density in high-altitude areas. Furthermore, SMI maps show their usefulness in this study before and after the flood storm and we will discuss them thereafter.

Exploiting the scatter plots between LST and NDVI, before (**Figure 8a**) and after (**Figure 8b**) the flood, the necessary coefficients (*a1, a2, b1,* and *b2*) for SMI (Eq. (7)) were obtained using linear regression for both dry and wet edges (Eqs. (8) and (9)). Then, the final SMI equations were established and implemented using OLI and TIRS data acquired before and after flood storm (SMI-1, Eq. (10)) and (SMI-2, Eq. (11)).

$$SMI\_1 = \frac{-17.00\ \text{NDVI} + 308.21\ -\ \text{LST}}{-37.68\ \text{NDVI}\ + 23.72} \tag{10}$$

$$SMI\_2 = \frac{-22.64.\text{NDVI} + 305.88 \quad - \quad \text{LST}}{-39.93.\text{NDVI} \quad + 24.39} \tag{11}$$

Figure 9 illustrates the SMI map before the flash flood, named SMI-1, highlighting the water content variability over the study site in different classes. High SM (> 60%) value classes are observed in areas located at the top of the mountains, surrounding Guelmim city, where vegetation cover is relatively dense. Also, significant SM content (~35%) characterizes small agricultural fields and oases with scattered vegetation cover, hydrographic network, watershed, and ravines. In the Guelmim plain, SMI values are very low and close to 0. In general, these lower SMI values indicated heavier drought occurrence, which is coherent with SMOS SM values over the study site during the same day. Furthermore, we observe that SMI-1 based on LST is sensitive to rock thermal properties, soil surface composition (roughness and mineralogy), soil hydraulic properties (unseen water, capillarity, and evaporation) that typically occur in the subsurface, topographic variation, and steady state groundwater temperature with a spatially homogeneous temperature distribution. Therefore, these sensitivity variations highlight the SM as a function of the geomorphology and the topography on the terrain. Statistical fit between several homologous points extracted from SMI-1 and the SM-RFE database shows significant correlation (R^2 of 0.90). Furthermore, **Figure 10** illustrates the derived SMI map, named SMI-2, from the OLI/TIRS data acquired on the 9th of December, i.e., after several days of flooding that occurred between November 21st and December 1st. It describes different SM classes following the subwatersheds, topographic variations, rock properties, and soil composition. Highest SM values (>75%) are detected in high altitude (read color class) associated with dense vegetation cover. However, in the south and southeast without vegetation cover, the high SM values are related to the rock types forming

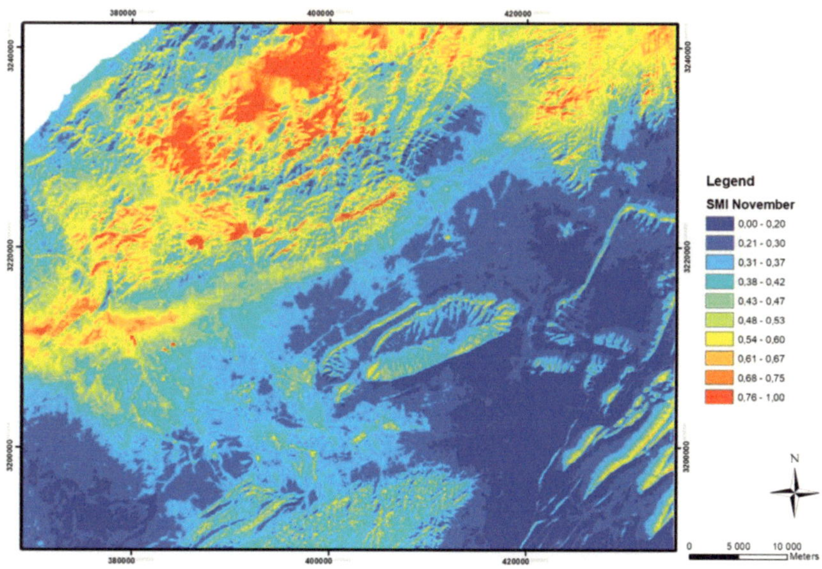

Figure 9. SMI-1 before the flood storm.

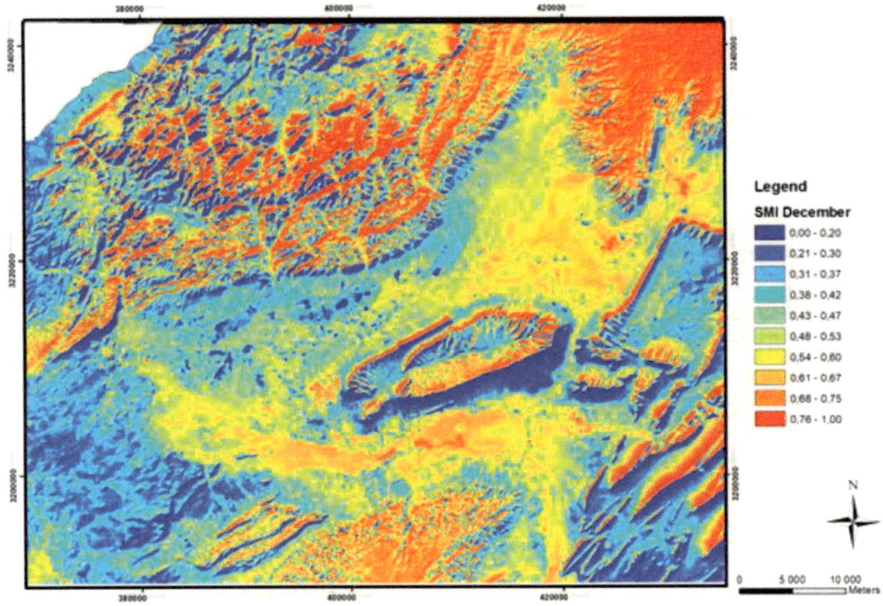

Figure 10. SMI-2 after the flood storm.

a synclinal located at high altitudes. In areas where water flow coming from the mountains following narrow streams via dejection cones, as well as the wadis (rivers) and their tributaries, was drained by gravity toward the plain of Guelmim where water and mud are accumulated forming a vast pool (E-W direction), the calculated SM values are very significant (45% ≤ SM ≤ 70%), i.e., green, yellow, and brown color classes. Areas of this class are covered by low permeable Quaternary carbonates and fluviolacustrine silts, allowing the water retention. Finally, the areas with relative medium slopes (9.0° to 20°) exhibit SM values between 30% and 42% (blue-cyan color class), while low SM values (<10%) are identified in areas with steep slopes facilitating water transition (blue color class). These highest SM values detected by SMI-2, 15 days after the flood storm, testify the strength of this tempest that killed 46 people and severely damaged the region's infrastructure (**Figure 1**). These results are in agreement with those retrieved from SMOS, and its validation against SM-RFE database also shows a significant correlation (R^2 of 0.95).

4. Conclusions

Flood-storm forecasting requires data collection, preprocessing, processing, and derived product interpretation to estimate the amount of runoff that will occur spatially and temporally. Then, the forecaster can identify when and where problems could occur for natural disaster management, and people assistance effort coordination during flooding periods. Among others, monitoring SM is a fundamental key for decision makers to predict and to manage flood event before, during, and after flash-flood storm. To meet these concerns, the past

10 years have witnessed the emergence of potentially transformative new remote sensing technologies and methods for SM estimation, which are beginning to fundamentally alter the possibilities for water-cycle and flood risk monitoring at different scales. The aim of this chapter is SM characterization over the Guelmim city and its neighborhood in the Southwestern Morocco that has been flooded several times. To achieve these, space-borne SMOS and Landsat-8 OLI/TIRS data were used. The SMOS BT data acquired before, during, and after the storm with 1-day temporal resolution and coarse spatial resolution (25 km) were preprocessed, processed, and transformed to the SM maps. OLI and TIRS data acquired 2 weeks before and 2 weeks after flood storm were preprocessed and transformed to NDVI and LST, and then combined to retrieve the SMI maps. The results obtained were analyzed and validated against the compiled SM values from rainfall data (SM-RFE) delivered by NOAA climate prediction center RFE for Africa.

SMOS results show how the spatial variation of SM changes extremely before, during, and after the flash flood with excellent temporal resolution over a very large territory (or at a country scale) independently to the meteorological conditions. It is possible to predict day-by-day the power, the cycle, and the direction of the storm progress. Obviously, this information is very useful for a warning system development for flash-flood prediction. By reference to the measured SM-RFE database considered as a ground truth, the validation of the derived SMOS SM maps exhibit a significant correlation ($R^2 > 0.90$). However, despite these significant and positive potentials, the major disadvantage of SMOS data for several hydrological applications, especially flash-flood storm monitoring, is unfortunately the coarse spatial resolution that is not sufficient for detailed and accurate information extraction locally or regionally at the large or medium scales. Furthermore, although the SM maps have been derived from OLI and TIRS based on different wavelengths, unlike acquisition modes, and using different image-processing methods, the results converge toward the same conclusions. The MSI results show a drastic SM change before and after flash flood, highlighting with more details the impact of inundation and the mud accumulation in the study site. Based on NDVI and LST, the SMI-1 shows certain sensitivity to the soil and rock thermal properties, the soil surface composition, the soil hydraulic properties, etc. Moreover, these variations highlight the SM as a function of the geomorphology and the topography on the terrain. For validation purposes, statistical fit between several homologous points extracted from SMI-1 and SM-RFE database shows a significant correlation (R^2 of 0.90). Likewise, 15 days after the flood storm with a saturated soil moisture, SMI-2 describes SM correctly. Indeed, significant correlation was obtained (R^2 of 0.95) when validating the derived SMI-2 map against SM-RFE datasets.

In this chapter, although the SM maps have been derived from data acquired with different sensors and using different image-processing methods, the results converge toward the same conclusions, before and after the flood storm. However, even if the spatial resolution of SMOS observations is adequate for several global applications, it is restricted to regional and local studies for flash-flood storm monitoring that requires medium spatial resolution. On the other hand, despite the potential of Landsat-8 (OLI and TIRS) medium scale data, they remain sensitive to other environmental factors especially meteorological conditions, which make the SM retrieval impractical during the storm. Nevertheless, the synergy between coarse spatial resolution microwave data and medium resolution optical-thermal data is likely helpful to achieve a multiresolution SM retrieval approach as previously discussed. Their integration in a GIS environment with other ancillary data will contribute significantly in the development of a prediction and monitoring model for flash flooding at the regional and local scales.

Acknowledgements

The authors would like to thank the Arabian Gulf University (AGU) for their financial support. We would like to thank the NASA-GLOVIS-GATE for the Landsat-8 (OLI/TIRS) data, the LP-DAAC NASA-USGS for the STM GDEM data, and the ESA for SMOS data. Our gratitude goes to many people who have made the used photos available on the web for consultation and public use. Finally, we express gratitude to the editors of this book for their constructive comments.

Author details

Abderrazak Bannari[1]*, Hassan Rhinane[2] and Hicham Bahi[3]

*Address all correspondence to: abannari@agu.edu.bh

1 Department of Geoinformatics, College of Graduate Studies, Arabian Gulf University, Manama, Kingdom of Bahrain

2 Department of Geology, Faculty of Sciences Ain Chock, University Hassan II, Casablanca, Morocco

3 Department of Computer Sciences, Faculty of Sciences Ain Chock, University Hassan II, Casablanca, Morocco

References

[1] IGWCO. Integrated Global Water Cycle Observation: Community of Practice. Earth Observation Group [Internet]. 2010. 20 p. Available from: http://www.wmo.int/pages/prog/hwrp/chy/chy14/documents/ms/IGWCO.pdf [Accessed: Jan 10, 2018]

[2] Moran MS, Peters-Lidard CD, Watts JM, McElroy S. Estimating soil moisture at the watershed scale with satellite-based radar and land surface models. Canadian Journal of Remote Sensing. 2004;**30**(5):805-826. DOI: 10.5589/m04-043

[3] Bannari A, Ghadeer A, El-Battay A, Hameid NA, Rouai M. Assessment of land erosion and sediment accumulation caused by runoff after a flash-flooding storm using topographic profiles and spectral indices. Advances in Remote Sensing. 2016;**5**:315-354. DOI: 10.4236/ars.2016.54024

[4] Bannari A, Kadhem G, El-Battay A, Hameed NA, Rouai M. Detection of areas associated with flash floods and erosion caused by rainfall storm using topographic attributes, hydrologic indices, and GIS. In: Pirasteh S, Li J, editors. Global Changes and Natural Disaster Management: Geo-Information Technologies. Springer International Publishing: Germany; 2017. pp. 155-174. DOI: 10.1007/978-3-319-51844-2.chp.13

[5] Joyce K, Belliss S, Samsonov S, McNeill S, Glassey P. A review of the status of satellite remote sensing and image processing techniques for mapping natural hazards and disasters. Progress in Physical Geography. 2009;**33**:183-207. DOI: 10.1177/0309133309339563

[6] Friedl MA, Davis FW. Sources of variation in radiometric surface temperature over a tallgrass prairie. Remote Sensing of Environment. 1994;**48**(1):1-17. DOI: 10.1016/0034-4257(94)90109-0

[7] Njoku E, Jackson E, Lakshmi V, Chan T, Nghiem S. Soil moisture retrieval from AMSR-E. IEEE Transactions on Geoscience and Remote Sensing. 2003;**41**:215-229. DOI: 10.1109/TGRS.2002.808243

[8] Ulaby FT, Dubois PC, Van Zyl J. Radar mapping of surface soil moisture. Journal of Hydrology. 1996;**184**:57-84. DOI: 10.1016/0022-1694(95)02968-0

[9] Kerr YH, Waldteufel P, Wigneron JP, Martinuzzi JM, Font J, Berger M. Soil moisture retrieval from space: The soil moisture and ocean salinity (SMOS) mission. IEEE Transactions on Geoscience and Remote Sensing. 2001;**39**(8):1729-1735. DOI: 10.1109/36.942551

[10] Bannari A, El-Harti A, Haboudane D, Bachaoui M, El-Ghmari A. Intégration des variables spectrales et géomorphométriques dans un SIG pour la cartographie des zones exposées à l'érosion. Revue Télédétection. 2008;**7**:393-404

[11] Maimouni S, Bannari A, El-Hrati A, El-Ghmari A. Potentiels et limites des indices spectraux pour caractériser la dégradation des sols en milieu semi-aride. Journal Canadien de Télédétection. 2012;**37**:285-301. DOI: 10.5589/m11-038

[12] Erskine WD, Saynor MJ. Effects of catastrophic floods on sediment yields in Southeastern Australia. Erosion and sediment yield: Global and regional perspectives. In: Proceedings of the Exeter Symposium, Exeter, UK. July 15–19, 1996. pp. 381-388

[13] SIGMA. Natural catastrophes and man-made disasters in 2014: Convective and winter storms generate most losses. Swiss Re Sigma [Internet]. 2015;**2**(50):1-47. Available from http://www.biztositasiszemle.hu/files/201503/sigma2_2015_en.pdf [Accessed: Mar 18, 2016]

[14] Alexakis DD, Mexis FPK, Vozinaki AEK, Daliakopoulos IN, Tsanis IK. Soil moisture content estimation based on Sentinel-1 and auxiliary earth observation products. A hydrological approach. Sensors. 2017;**17**(6):1455. DOI: 10.3390/s17061455

[15] Wengler L, Weisrock A, Brochier JE, Brugal JP, Fontugne M, Magnin F, et al. Enregistrement fluviatile et paleo-environnements au Pleistocene superieur sur la bordure meridionale

atlantique de l'Anti-Atlas (Oued Assaka, S-O marocain). Quaternaire. 2002;**13**:179-192. DOI: 10.3406/quate.2002.1710

[16] Kerr YH, Waldteufel P, Richaume P, Wigneron JP, Ferrazzoli P, Mahmoodi A, et al. The SMOS soil moisture retrieval algorithm. IEEE Transactions on Geoscience and Remote Sensing. 2012;**50**(5):1384-1403. DOI: 10.1109/TGRS.2012.2184548

[17] Kerr YH, Jacquette E, Al-Bitar A, Cabot F, Mialon A, Richaume P, et al. CATDS SMOS L3 soil moisture retrieval processor algorithm theoretical baseline document (ATBD). Report of CNES and CATDS, France. Reference: SO-TN-CBSA-GS-0029. 2013. 73 p

[18] Bannari A, Teillet PM, Landry R. Comparaison des réflectances des surfaces naturelles dans les bandes spectrales homologues des capteurs TM de Landsat-5 et TME+ de Landsat-7. Revue Télédétection. 2004;**4**:263-275

[19] Wulder MA, White JC, Masek JG, Dwyer J, Roy DP. Continuity of Landsat observations: Short term considerations. Remote Sensing of Environment. 2012;**115**:747-751. DOI: 10.1016/j.rse.2010.11.002

[20] USGS. Shuttle Radar Topography Mission [Internet]. 2008. Available from: http://srtm.usgs.gov/Mission/missionsummary.php [Accessed: Jun 10, 2015]

[21] NASA. Releases Enhanced Shuttle Land Elevation Data [Internet]. 2015. Available from: http://www2.jpl.nasa.gov/srtm/ [Accessed: Jan 10, 2016]

[22] Bannari A, Kadhem G, El-Battay A, Hameid N. Comparison of SRTM-V4.1 and ASTER-V2.1 for accurate topographic attributes and hydrologic indices extraction in flooded areas. Journal of Earth Science and Engineering. 2018;**8**:8-30. DOI: 10.17265/2159-581X/2018.01.002

[23] Maidment RI, Grimes D, Black E, Tarnavsky E, Young M, Greatrex H, et al. A new, long-term daily satellite-based rainfall dataset for operational monitoring in Africa. Scientific Data. 2017;**4**:170063. DOI: 10.1038/sdata.2017.63

[24] NOAA. Climate Prediction Center (CPC) Rainfall Estimator (RFE) for Africa [Internet]. 2015. Available from: https://catalog.data.gov/dataset/climate-prediction-center-cpc--rainfall-estimator-rfe-for-africa [Accessed: Feb 10, 2018]

[25] Lam A, Bierkens MFP, Van-den-Hurk BJJM. Global patterns of relations between soil moisture and rainfall occurrence in ERA-40. Journal of Geophysics Research. 2007;**112**: D17116. DOI: 10.1029/2006JD008222

[26] Wigneron JP, Calvet JC, De-Rosnay P, Kerr YH, Waldteufel P, Saleh K, et al. Soil moisture retrievals from bi-angular L-band passive microwave observations. IEEE Transactions on Geoscience and Remote Sensing. 2004;**1**:277-281. DOI: 10.1109/LGRS.2004.834594

[27] Lievens H, Tomer SK, Al Bitar A, De Lannoy GJM, Drusch M, Dumedah G, et al. SMOS soil moisture assimilation for improved hydrologic simulation in the Murray Darling Basin, Australia. Remote Sensing of Environment. 2015;**168**:146-162. DOI: 10.1016/j.rse.2015.06.025

[28] Myneni RB, Asrar G. Atmospheric effects and spectral vegetation indices. Remote Sensing of Environment. 1994;**17**:390-402. DOI: 10.1016/0034-4257(94)90106-6

[29] Bannari A, Teillet PM, Richardson G. Nécessité de l'étalonnage radiométrique et standardisation des données de télédétection. Canadian Journal of Remote Sensing. 1999; **25**:45-59. DOI: 10.1080/07038992.1999.10855262

[30] Teillet PM, Santer RP. Terrain elevation and sensor altitude dependence in semi-analytical atmospheric code. Canadian Journal of Remote Sensing. 1991;**17**:36-44

[31] Bannari A, Morin D, Bénié GB, Bonn F. A theoretical review of different mathematical models of geometric corrections applied to remote sensing images. Remote Sensing Reviews. 1995; **13**:27-47. DOI: 10.1080/02757259509532295

[32] USGS. Using the USGS Landsat-8 Product [Internet]. 2018. Available from: https://landsat.usgs.gov/using-usgs-landsat-8-product [Accessed: Feb 10, 2018]

[33] Sobrino JA, Jimenez-Muoz JC, Soria G, Romaguera M, Guanter L, Moreno J, et al. Land surface emissivity retrieval from different VNIR and TIR sensors. IEEE Transactions on Geoscience and Remote Sensing. 2008;**46**:316-327. DOI: 10.1109/TGRS.2007.904834

[34] Wigneron JP, Jackson TJ, O'Neill P, De Lannoy G, De Rosnay P, Walker JP, et al. Modelling the passive microwave signature from land surfaces: A review of recent results and application to the L-band SMOS and SMAP soil moisture retrieval algorithms. Remote Sensing of Environment. 2017;**192**:238-262. DOI: 10.1016/j.rse.2017.01.024

[35] Kerr YH, Al-Yaari A, Rodriguez-Fernandez N, Parrens M, Molero B, Leroux D, et al. Overview of SMOS performance in terms of global soil moisture monitoring after six years in operation. Remote Sensing of Environment. 2016;**180**:40-63. DOI: 10.1016/j.rse.2016.02.042

[36] Fernandez-Moran R, Wigneron J-P, De Lannoy G, Lopez-Baeza E, Parrens M, Mialon A, et al. A new calibration of the effective scattering albedo and soil roughness parameters in the SMOS SM retrieval algorithm. International Journal of Applied Earth Observation and Geoinformation. 2017;**62**:27-38

[37] Escorihuela MJ, Chanzy A, Wigneron JP, Kerr YH. Effective soil moisture sampling depth of L-band radiometry: A case study. Remote Sensing of Environment. 2010;**114**(5):995-1001. DOI: 10.1016/j.rse.2009.12.011

[38] Zeng Y, Feng Z, Xiang N. Assessment of soil moisture using Landsat ETM+ temperature/vegetation index in semiarid environment. In: Proceedings of International Geoscience and Remote Sensing Symposium, IGARSS '04; Anchorage, AK, USA. Vol. 6. Sep 20–24, 2004. pp. 4306-4309

[39] Rodrıguez-Fernandez NJ, Aires F, Richaume P, Kerr YH, Prigent C, Kolassa J, et al. Soil moisture retrieval using neural networks: Application to SMOS. IEEE Transactions on Geosciences and Remote Sensing. 2015;**53**(11):5991-6007. DOI: 10.1109/TGRS.2015.2430845

[40] Chen Y, Yang K, Qin J, Cui Q, Lu H, La Z, et al. Evaluation of SMAP, SMOS, and AMSR2 soil moisture retrievals against observations from two networks on the Tibetan Plateau.

Journal of Geophysical Research—Atmosphere. 2017;**122**:5780-5792. DOI: 10.1002/2016JD026388

[41] Tramblay Y, Bouvier C, Martin C, Didon-Lescot JF, Todorovik D, Domergue JM. Assessment of initial soil moisture conditions for event-based rainfall-runoff modelling. Journal of Hydrology. 2010;**387**(3–4):176-187. DOI: 10.1016/j.jhydrol.2010.04.006

[42] Wooldridge SA, Kalma JD, Walker JP. Importance of soil moisture measurements for inferring parameters in hydrologic models of low-yielding ephemeral catchments. Environmental Modelling and Software. 2003;**18**(1):35-48. DOI: 10.1016/S1364-8152(02)00038-5

[43] Aubert D, Loumagne C, Oudin L. Sequential assimilation of soil moisture and streamflow data in a conceptual rainfall-runoff model. Journal of Hydrology. 2003;**280**(1–4):145-161. DOI: 10.1016/S0022-1694(03)00229-4

[44] Massari C, Brocca L, Tarpanelli A, Moramarco T. Data assimilation of satellite soil moisture into rainfall-runoff modelling: A complex recipe? Remote Sensing. 2015;**7**(9):11403-11433. DOI: 10.3390/rs70911403

[45] Cenci L, Laiolo P, Gabellani S, Campo L, Silvestro F, Delogu F, et al. Assimilation of H-SAF soil moisture products for flash flood early warning systems. Case study: Mediterranean catchments. IEEE Journal of Selected Topics in Applied Earth Observation Remote Sensing. 2016;**9**:5634-5646. DOI: 10.1109/JSTARS.2016.2598475

[46] Alvarez-Garreton C, Ryu D, Western AW, Su CH, Crow WT, Robertson DE, et al. Improving operational flood ensemble prediction by the assimilation of satellite soil moisture: Comparison between lumped and semi-distributed schemes. Hydrology and Earth System Sciences. 2015;**19**(4):1659-1676. DOI: 10.5194/hess-19-1659-2015

[47] Brocca L, Ciabatta L, Massari C, Camici S, Tarpanelli A. Soil moisture for hydrological applications: Open questions and new opportunities. Water. 2017;**9**:140. DOI: 10.3390/w9020140

Chapter 3

Development and Application of Conceptual Rainfall-Altitude Regression Model: The Case of Matahara Area (Ethiopia)

Megersa Olumana Dinka

Additional information is available at the end of the chapter

http://dx.doi.org/10.5772/intechopen.80694

Abstract

Rainfall data available, in tropical regions with undulating topography, may provide a valuable information for water resource development as well as for predicting and preventing natural disasters. But in developing countries like Ethiopia, rain gauge stations are sparsely populated, and rainfall data are the limiting factor. Hence, estimation of rainfall is extremely important. The current paper deals with the development of a rainfall-altitude relationship for Matahara area, Awash Basin of Ethiopia. A conceptual rainfall-altitude regression model was formulated and its performance evaluated. The relationship between monthly rainfall totals and gauge elevation over Matahara region (including Lake Basaka catchment) was examined using the conceptual regression model (ordinary least square). The regression parameters were identified and estimated and then used to map the spatial rainfall for Lake Basaka catchment in ArcGIS. The regression analysis showed a strong positive correlation ($r = 0.85$) between the long-term average monthly rainfall and altitude of the region. It is shown that the rate of increase of rainfall with altitude is in the range of 0.020 mm/h at Matahara to 0.067 at Welenchiti, with average value of 0.0475 mm/m/month. The best fit ($R^2 = 0.9187$, $p = 0.015$) was obtained between observed and estimated rainfall depths for all the stations with total standard error of 12.97 mm. The high R^2 reveals that the developed equation is acceptable for the area at 98.5% ($p > 0.015$) confidence limit. The performance of the developed model is found to be within reasonable accuracy, which is limited by the elevation difference and distance from the base station. Therefore, the spatial and temporal structures of rainfall distribution (daily, monthly or annual) for Matahara region (including Basaka Lake catchment) can be determined from the available records of rainfall data at the Matahara Research Station (Merti) meteorological station with acceptable reliability. In general, the performance of the developed model is found to be within reasonable accuracy, which is limited by the elevation difference and distance from the base station.

Keywords: Matahara, orography, performance, rainfall, regression model

1. Introduction

Accurate estimates of the spatiotemporal distribution of rainfall have been found to play a key role in hydrological applications and water resource management [1]. Tropical regions with heavy rainfall and variable topography are characterized by serious natural disasters. Rainfall estimates, in such region, provide valuable information for water resources development and for predicting and preventing natural disasters [2].

Rainfall shows variations in amount and distribution depending upon the factors: wind (speed and direction), topography (altitude and slope), barrier characteristics, mountain scales, etc. (e.g. [2–4]). Although the seasonal and spatial distribution of the rainfall is the key parameter for further differentiation of climate, orography is the crucial parameter for the development of the regional climatic behavior in Ethiopia [3, 5]. It is evident from the previous investigations that rainfall increases with altitude (e.g. [1, 5]). However, understanding the nature of rainfall-elevation relationships in mountainous regions is much difficult.

Matahara area, including Lake Basaka catchment, has a wide range of topographic elevations, ranging from 949 m at the Lake to over 1900 m at the top of Fentalle Mountain [4]. Hence, a spatial rainfall variation in the region is expected. The reliable estimate of rainfall spatial distribution is extremely important for the determination of the regimes of hydrologic processes (runoff, erosion, sedimentation) within the lake catchment and the resulting regime of lake's water balance. No profound study has been made so far on the stochastic model of rainfall-altitude relationship in the study area.

In this study, the topographic effects on rainfall distribution has been investigated and explained. An attempt was made to determine the spatial distribution of the rainfall from the available rainfall recording stations within the region with similar agroclimatic condition and variable altitude coverage. A conceptual rainfall-elevation regression model was formulated and then simulated for Matahara region in order to see its practicability for the area. The regression model formulation is based on the hypothesis: rainfall at certain location is the function of rainfall data measured at the base station, the elevation difference between the two locations, and the incremental rainfall per altitude.

2. Materials and methods

2.1. Study area

Matahara area is situated within Awash River Basin, central rift valley of Ethiopia. It is located within Oromia region. Matahara plain area, in general, has semiarid climate [4], with a mean annual rainfall of about 543.7 mm (**Figure 1**). As evident from **Figure 1**, the plain area is characterized by bimodal and erratic rainfall distribution. Details of Matahara area including Lake Basaka are well documented [4, 6–9]. The study area is attracting the attention of many scholars due to the fact that the highly expanding Lake Basaka is situated within the region. The lake is expanding at very fast rate in the past about 5 decades. The main problem with

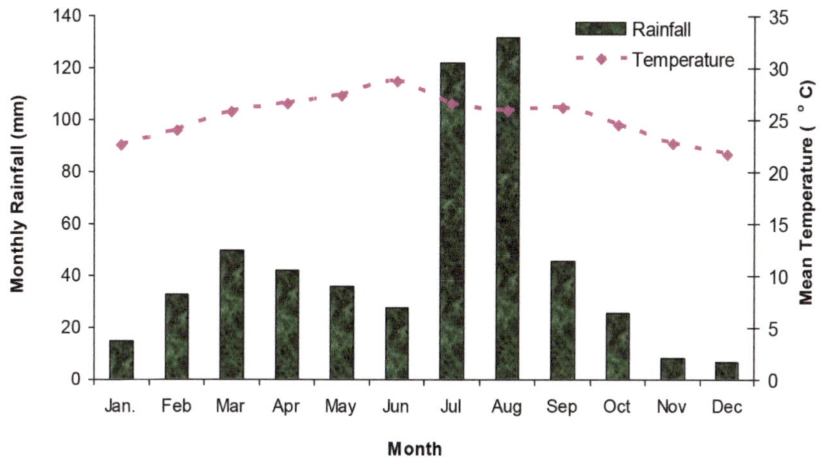

Figure 1. Mean average seasonal variability of rainfall and temperature in Matahara area.

the lake expansion is due to its poor water quality (high salinity, sodicity and alkalinity), and it is not usable for domestic or irrigation purpose. Moreover, the lake is also situated within the main rift valley of Ethiopia (at a close distance to Afar Triangle), where many changes are happening.

2.2. Rainfall estimation

There are two meteorological stations in Matahara region with long years of recorded data: one at Matahara Breeding Station (MSF) and one at Matahara town just near to the lake (**Figure 2**). Unfortunately both stations are outside the Lake catchment and even cannot be representative for the entire catchment since they are located almost at the lower elevation. There are also other two meteorological stations (Awash and Nura-Era) in the vicinity of the Lake. Though the second station is being located at the southern most part within the Lake catchment, it cannot be representative for the entire catchment too. Furthermore, there are a number of rain gauges (about 12) distributed in the sugar plantation section with 12-year (1996–2008) rainfall records, 4 of which are located in the lake catchment from Abadir side. Even these rain gauges are concentrated at one location (with low elevation), and their measurements are less reliable.

Therefore, an attempt was made to determine the spatial rainfall distribution for the area from the available rainfall recording stations within the region with similar agroclimatic condition (semiarid) and variable altitude coverage. About seven rain gauge stations with long year's measurement data were selected, namely, Matahara Research Station (MRS) at Merti, Matahara, Nura-Era, Awash, Welenchiti, Adama and Wonji Research Station (WRS) (**Figure 2**). These stations have good-quality, continuous records of data for the period of 1966–2010. As confirmed by quality check, the recorded data of the seven stations are homogeneous (at the >95% confidence level) and consistent. Data records of Welenchiti station are found to be relatively poor compared to the others.

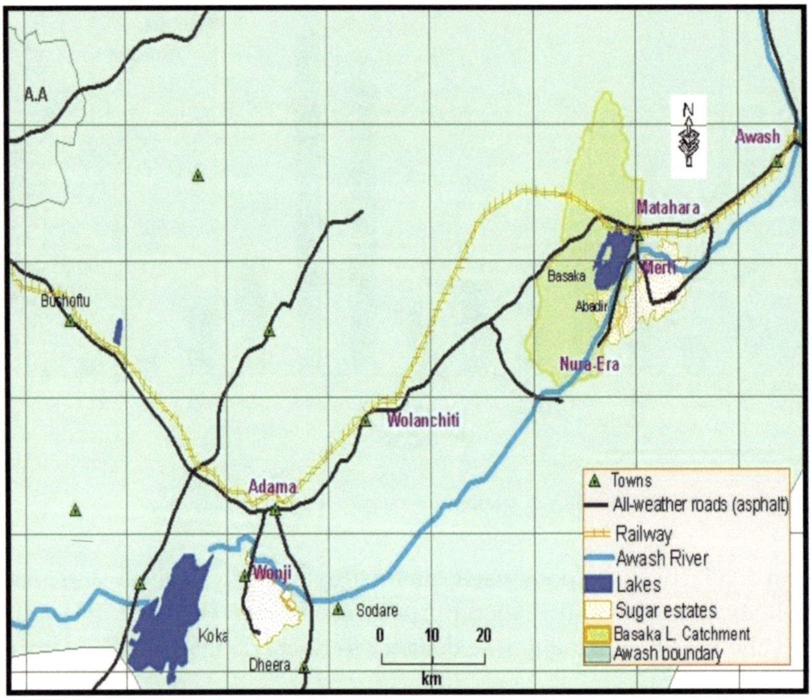

Figure 2. Distribution of meteorological stations considered in the region.

2.3. Formulation of the regression model

Most studies conducted in the mid-latitudes have proved that a simple linear rainfall-altitude regression model well fits the observed data [10]. Accordingly, a conceptual regression model (Eq. (1)) was first formulated based on the hypothesis that rainfall at certain location is the function of rainfall data measured at the base station, the elevation difference between the two locations, and the incremental rainfall per altitude. The simple linear regression model was fitted by ordinary least squares (OLS) method:

$$P_H = P_b + k * (H - H_b) \pm \varepsilon \tag{1}$$

where P (mm) = rainfall at elevation H, P_b = rainfall at base station H_b, H (m) = elevation at which P is to be determined, H_b (m) = elevation of the base station, k (mm/m) = constant as function of rainfall increment per altitude and ε = error term that considers the effect of error in measurement and effect of other factors on rainfall. The rainfall increment per altitude (k) value is different for monthly and annual rainfall values. The expectation from the regression model is that the rainfall increases with altitude, i.e., k should be positive.

The above equation (Eq. (1)) was optimized until the computed and measured rainfalls (P) are approximately equal or until the sum of the deviations between the observed and computed

http://dx.doi.org/10.5772/intechopen.80694

rainfalls are approximately zero (standard error). For the lake catchment (Matahara area), the meteorological station at MRS was considered as the base station since it has long years of measurement data and its records are found to be relatively reliable and consistent (as checked by double mass curve) and is also at close vicinity to the lake and its catchment.

3. Results and discussion

3.1. Rainfall-altitude relationship

The temporal variability of long-term (1966–2008) mean monthly rainfall values for the selected meteorological stations (excluding Adama and Awash Melkassa) is plotted in **Figure 3**. The regression analysis showed a strong positive correlation (r = 0.85) between the LYA monthly rainfall and altitude of the region (**Figure 4**). The correlation becomes even stronger (r = 0.97)

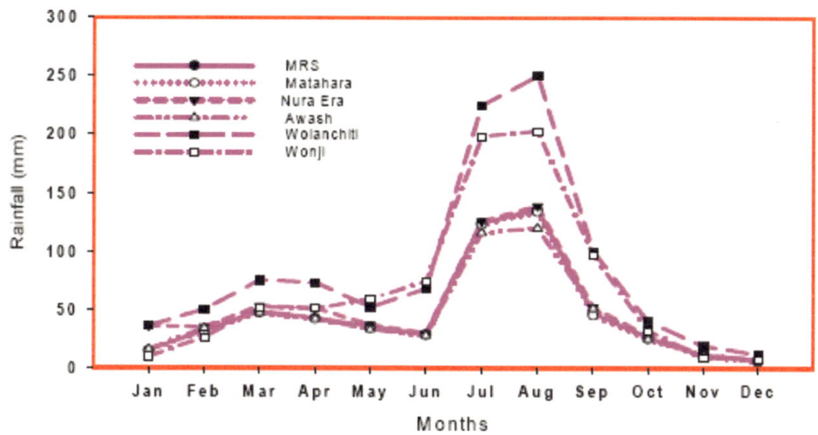

Figure 3. Temporal variability of LYA annual rainfall of Awash valley at six recording stations.

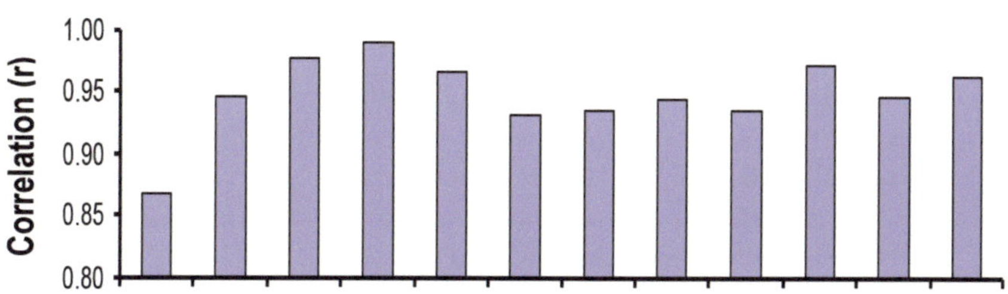

Month	Jan	Feb	Mar	Apr	May	Jun	Jul	Aug	Sep	Oct	Nov	Dec	Annual
r	0.869	0.948	0.978	0.991	0.967	0.932	0.936	0.945	0.936	0.972	0.948	0.964	0.970

Figure 4. Correlation analysis between rainfall and altitude (for all considered stations).

between the long years' average annual rainfall and the elevation, which reveals how topography (orography) is the most decisive factor for rainfall pattern of the region, which is also true for most part of Ethiopia.

The rainfall of the area is highly erratic and of bimodal type with the main rainy season concentrated to few months (July–September). Analysis of the seasonal rainfall distribution pattern showed that the most intensive hydrologic processes (runoff, erosion and sedimentation) in the area are expected in the month of July and/or August, which was also confirmed by other studies [4, 9]. It possible to suggest that the irregular rainfall patterns of the area may be related to the cyclic events of climate such as QBO and ENSO in the tropics [11]. African rainfall variability is mostly related to the ENSO indices [11, 12]. Like tropics, Southern Africa regions have high climate variation and hence susceptible to droughts and floods [13–15].

3.2. Developed conceptual regression model

A conceptual regression model was formulated (see Eq. (1)) based on the correlation obtained between rainfall and altitude of the region (**Figure 4**). The conceptual regression model was optimized until the computed and measured rainfalls (P) are approximately equal or until the sum of the deviations between the observed and computed rainfall are approximately zero (standard error). After optimization, the regression model (Eqs. (2) and (3)) was developed for the area. After inserting the elevation value (H_b = 956 m) for the base station (MRS), Eq. (2) is reduced to Eq. (3):

$$\begin{aligned} P(H) &= P_b + 0.0475 * (H - H_b) - 0.9 \quad (monthly) \\ P(H) &= P_b + 0.560 * (H - H_b) - 0.9 \quad (Annual) \end{aligned} \tag{2}$$

$$P = P_b + 0.0475 * H - 530.8 \tag{3}$$

The incremental rainfall per altitude obtained for the considered seven stations is in the range of 0.020 (Matahara) to 0.067 (Welenchiti), with average value of 0.0475 mm/m/month. The monthly rainfall of each of the considered stations was computed using Eq. (3) and compared to the observed values (**Figure 5**). The performance of the developed regression model was evaluated based on the RMSE (**Table 1**). Welenchiti and Wonji have shown the highest RMSE of 70 and 120 mm, respectively. Conversely, the other four stations (MRS, Matahara, Awash and Nura-Era stations) in the close vicinity to the lake have the lowest RMSE, indicating the better fit of data. That means those stations very far (distance > 50 km) and with high elevation difference (>200 m) from the base station perform poor.

The scatter plot of the observed and simulated rainfall for the considered stations in the region is shown in **Figure 6**. The best fit (R^2 = 0.9187, p = 0.015) was obtained between observed and estimated rainfall depths for all the stations (**Figure 6b**) with total standard error of 12.97 mm. The correlation (R^2 = 0.9988) is very strong for the selected four stations (MSE, Matahara, Awash and Nura-Era) in the vicinity of the lake (**Figure 6a**). The high R^2 reveals that the developed equation is acceptable for the area at 98.5% (p > 0.015) confidence limit. Therefore, the monthly or annual rainfall for the lake and its catchment can be estimated from

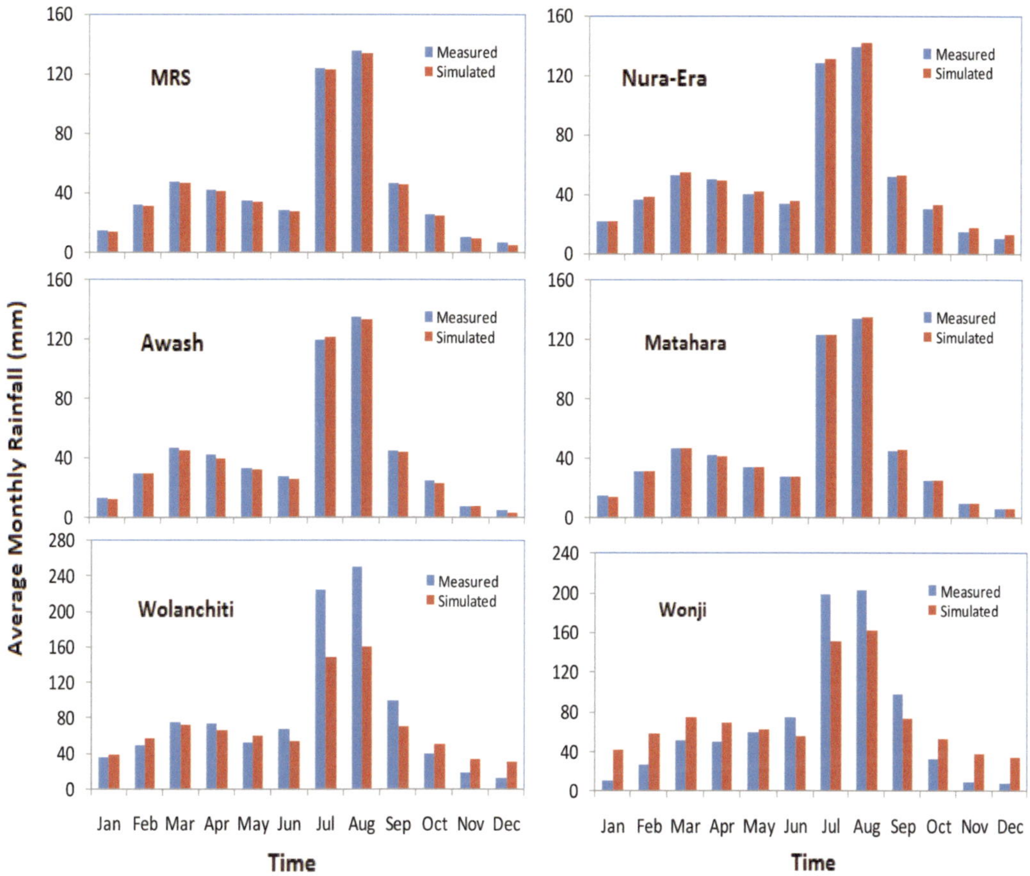

Figure 5. Mean long years' average monthly measured and simulated rainfall (1966–2008).

Station	RMSE (mm)	Elevation difference (m) from base station	Approximate distance from base station (km)
MRS	0.90	0	0
Matahara	0.20	−2	8
Awash	4.01	−30	30
Nura-Era	6.02	165	40
Wonji	70.00	550	100
Welenchiti	120.00	535	70

RMSE, root mean square error.

Table 1. Performance of the developed regression for the different stations.

the measured data at the MRS meteorological station with acceptable reliability. Of course, some discrepancy between the estimated and measured rainfall is expected due to certain factors: measurement error, wind oscillation effects, altitude difference in the watershed, etc.

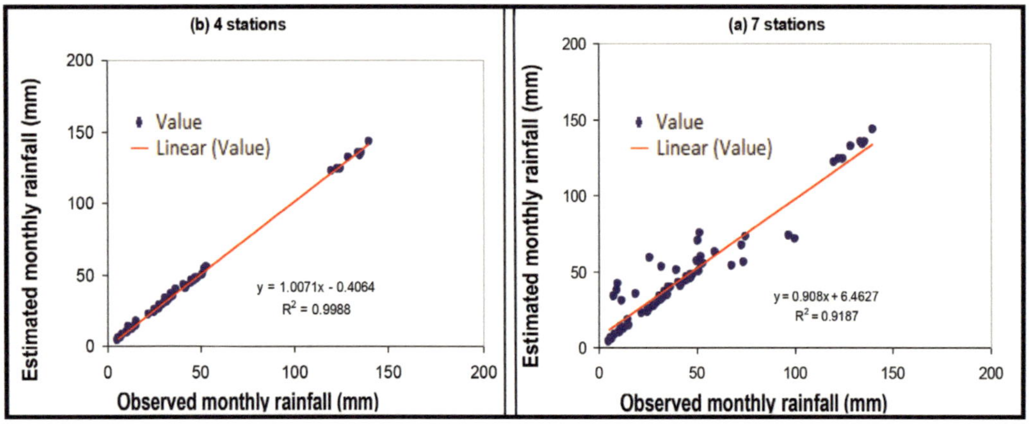

Figure 6. Scatter plot of observed and model estimated monthly rainfall depths: (a) for four stations in the vicinity of the lake and (b) for all stations considered.

4. Conclusion

The rainfall of the area is found to be erratic and of bimodal type, with great seasonal and annual variation. The seasonal variation is indicator of the occurrence of the most intensive hydrologic processes (runoff, erosion and sedimentation) in the months of July and August. Unlike rainfall, temperature showed an increasing trend. The mean average temperature increment observed for the area (+2.4°C) in the past 4–5 decades is slightly higher than that of the country's average and of the globe (+2°C) in the postindustrial period. This could be attributed to the massive deforestation and tectonism (volcanic activity). This change has significant implications on the hydrologic cycle of the area (local scale) and on the global warming (global scale).

Good correlation was observed between altitude and rainfall. The developed conceptual regression model (after optimization) showed that the incremental rainfall per altitude obtained for the eight stations in the region is in the range of 0.25–0.80, with average value of 0.56 mm/m/year. The model performance is in line with the formulated hypothesis and is found to be satisfactory for the region. However, the best performance was observed for four stations in the vicinity of the lake and its catchment. Therefore, the developed conceptual regression model can be applied for the computation of a spatial rainfall distribution of the area, including Lake Basaka catchment.

Author details

Megersa Olumana Dinka

Address all correspondence to: magarsol@yahoo.com

Department of Civil Engineering Sciences, Faculty of Engineering and the Built Environment, University of Johannesburg, Johannesburg, South Africa

References

[1] Suzuki Y, Nakakita E, Hasebe M, Ikebuchi S. Study on rainfall-topography relationships in Japan with regard to the spatial scale of mountain slopes. In: Sixth International Symposium on Hydrological Applications of Weather Radar; Melbourne, Australia. 2004

[2] Kuraji K, Punyatrong K, Suzuki M. Altitudinal increase in rainfall in the Mae Chaem Watershed, Thailand. Journal of the Meteorological Society of Japan. 2001;**79**(1B):353-363

[3] Gissila T, Black E, Grimes DIF, Slingo JM. Seasonal forecasting of the Ethiopian summer rains. International Journal of Climatology. 2004;**24**:1345-1358

[4] Dinka MO. Analyzing the extents of Basaka lake expansion and soil and water quality status of Matahara irrigation scheme, Awash Basin (Ethiopia) [PhD dissertation]. Vienna, Austria; BOKU University; 2010

[5] Seleshi Y, Zanke U. Recent changes in rainfall and rainy days in Ethiopia. International Journal of Climatology. 2004;**24**:973-983

[6] Abejehu G. Assessment of salinity & sodicity status of Matahara Sugar Estate [MSc thesis]. Ethiopia: Alemaya University of Agriculture; 1993

[7] Dinka MO, Loiskandl W, Furst J. The expansion of highly saline Basaka Lake and its effects on the sustainability of Matahara Sugar Estate. In: Proceedings of the 34th–WEDC International Conference; Addis Ababa, Ethiopia. Refereed Paper. 2009;**296**:571-579

[8] Dinka MO. Analysing the extents (size and shape) of Lake Basaka expansion (main Ethiopian Rift Valley) using remote sensing and GIS. Lake and Reservoirs: Research and Management. 2012;**17**:131-141

[9] Dinka MO, Loiskandl W, Ndambuki JM. Hydrochemical modelling for Lake Basaka: Development and application of a conceptual water budget model. Environmental Monitoring and Assessment. 2014;**186**:5365-5379

[10] Brunsdon C, McClatchey J, Unwin DJ. Spatial variations in the average rainfall–altitude relationship in greatbritain: An approach using geographically weighted regression. International Journal of Climatology. 2001;**21**:455-466

[11] Abtew W, Melesse A, Dessalegne T. Blue Nile basin hydrology relationship to climate indices. In: Proceedings of workshop on Hydrology and Ecology of the Nile River Basin under extreme conditions; Addis Ababa. 2008. pp. 76-89

[12] Jury MR. Intra-seasonal convective variability over southern Africa. Principal component analysis of pentad outgoing-longwave radiation departures. Theoretical and Applied Climatology. 1998;**62**:133-146

[13] Jury MR. An intercomparison of model-simulated east-west climate gradient over South Africa. Water SA. 2012;**38**:467-478

[14] Lyon B, Dewitt DG. A recent and abrupt decline in the east African long rains. Geophysical Research Letters. 2012;**39**:L02702

[15] Dedekind Z, Engelbrecht FA, van der Merwe J. Model simulations of rainfall over southern Africa and its eastern escarpment. Water SA. 2016;**42**(1):129-143

Chapter 4

Linkages between Water and Forests in South American Watersheds under Restoration

Denise Taffarello, Diego Alejandro Guzman Arias, Danielle de Almeida Bressiani, Davi Gasparini Fernandes Cunha, Maria do Carmo Calijuri and Eduardo Mario Mendiondo

Additional information is available at the end of the chapter

http://dx.doi.org/10.5772/intechopen.82526

Abstract

Water security is threatened by the rapid growth of the human population in areas where there were native forests before coupled with climate change scenarios. One of the main elements which ensures water security is water stored in soil, which is fundamental for maintaining ecohydrological processes at the watershed scale under forest land-use change. In South America, aiming to restore and recover changing catchment areas, best management practices (BMP) have been widely proposed as a strategy for water-forest resource sustainability. Based on forest evapotranspiration demand, this chapter presents fundamental concepts related to soil-water-forest cycles, watershed restoration, and case studies of BMPs in South American watersheds (e.g., Brazilian and Colombian projects for watershed conservation or restoration). It has become clear that there is an opportunity in setting baseline data and quantifying the effectiveness of these BMPs. By using ecohydrological monitoring and suitable indicators of these BMPs in the long term, an integrated understanding of water-forest relationships is needed. Furthermore, the more successful watershed management projects are, the more effective decision-making regarding BMP linking water and forests is.

Keywords: water yield, watershed restoration, hydrological services, ecohydrological processes, South America

1. Introduction

There is no life without water. Before the earliest writing systems evolved, humans were hunters and food gatherers. Then, with the onset of the Neolithic revolution, humans started settling alongside rivers and developed into early state civilizations, often referred to as "hydraulic civilizations." During the late Holocene, in Latin America, the Maya and Inca empires developed ancient water systems based on empirical observations. Urban civilization replaced small villages with towns and cities, and agriculture progressively took the place of native forests. However, new concerns arose in dealing with the multiple uses of water resources. On one hand, an evolution of public and industrial water supply systems was needed for human well-being. On the other hand, the development of such technologies, as well as other anthropogenic impacts, made ecosystems even more vulnerable than they already were.

Since the mid-twentieth century, rapid human population growth, technological development, and rising resource consumption have increased water pollution and scarcity. The *Food and Agriculture Organization* (FAO) of the United Nations [1] estimates that there will be a need to increase the global food production by 60% to feed more than 9 billion people foreseen to live in the world by 2050. The human impact on Earth seems to reflect a new period in the geological time scale: the so-called Anthropocene [2]. Waters et al. [3] summarized the key markers of functional changes due to anthropogenic actions, which are indicative of the Anthropocene, for example, biotic changes, which include species invasions and accelerated extinction. Furthermore, human-induced stressors are altering freshwater, marine, and terrestrial ecosystems in an unparallelled way [4].

The water cycle connects the abiotic environment with the bio- and anthropospheres, thereby leading the distribution of life on Earth [5]. In turn, freshwater ecosystems have been recognized among the most threatened ecosystems in the world from at least 20 years ago [6–9]. At least 10,000–20,000 freshwater species are already extinct or at risk, with loss rates comparable to those of the late Pleistocene–Holocene succession. Overexploitation and habitat loss trends are pushing Earth to the sixth mass extinction process [3]. It has been shown that 65% of Earth's river discharge and associated habitats are moderately to highly threatened [10].

In his pioneering work, Tansley [11] proposed the term "ecosystem," encompassing abiotic and biotic factors, as well as their functional and structural relationships. Moreover, terrestrial ecosystems influence freshwater by moving and modifying flows through a series of ecohydrological processes. The relationships among ecohydrological processes are strongly nonlinear (see [12–15]; also called geo-bio-hydrologic processes by [16]). This ecohydrological processes in both aquatic and terrestrial ecosystems provide benefits for humans, which are called ecosystem services [17, 18]. Various authors (i.e., [19–23]) have defined ecohydrological processes when relating ecological aspects of the hydrological cycle. For example, by the time correlation of the variable fraction of flooded areas with the duration of flood pulse can better integrate both the nutrient cycling and the river flow as part of a local biogeochemical cycle.

In the scope of this chapter, linked to water resources, three classes of ecosystem services, *provisioning*, *regulating*, and *supporting* and their links, are presented. All these types of ecosystem services have been progressively damaged by anthropogenic pressures:

- *Provisioning services* involve the production of renewable resources (e.g., freshwater, food, and extraction of pharmaceutical and cosmetic products from biota; [18]).
- *Regulating services* are those that indicate benefits arising from regulating ecological processes and, hence, lessen environmental change (e.g., climate regulation, water regulation throughout attenuation of hydrologic extremes such as floods and droughts, disease control; [18, 24]).
- *Supporting services* are the cycles of transformation of energy and mass at ecosystems, and they are the basis for providing other ecosystem services (such as nutrient cycles, soil formation; [18]).

Land-use/land-cover (LULC) changes due to anthropic activities are the main threats to the ecosystem regime shifts. Unbalanced water flows, biodiversity losses, and interruption of nitrogen, phosphorus, and other biogeochemical cycles deplete ecosystem services on large scales [17, 25]. Considering these significant changes, we need to understand how the ecohydrological processes work, if we want to develop better policies on watershed management [26].

The quantity and quality of water resources of each headwater are related to geology, topography, soil type, climate, type and amount of vegetal cover, and to the degree and type of anthropic activity in the watershed. Watershed restoration provides a variety of goods and services for humans and nature, including regulating water and ecosystem flows, improving water quality, reducing sediment loads, and affecting pollination and biodiversity.

The *ecosystem-based adaptation* (EbA) concept emerged at the beginning of the 2010 to mitigate the impacts of the climate change and anthropic activities. EbA means *the use of biodiversity and ecosystem services to help people adapt to the adverse effects of climate change*. This concept was defined by the Convention on Biological Diversity in the 10th Conference of the Parties [27]. Protecting ecosystem services is essential to promote watershed-scale sustainability to decrease ecosystems′ and people's vulnerability, as well as increase their resilience to global change impacts [28, 29]. The *watersheds restoration in South America* can be achieved through projects of payment for ecosystem services (PES) since PES projects are considered a method of EbA [30].

Neither integrated quali-quantitative analysis nor combined indicators of human-ecosystem appropriation of freshwater resources have been established in Brazilian basin plans [31, 32]. On the one hand, among these indicators, we highlight the water footprint—WF [33, 34]. It encompasses gray, blue, and green portions of water into a unique indicator to evaluate sustainability arising from water resource pollution and consumption. For sustainable water allocation planning, river plans must be built based on accurate data on actual water availability per basin, taking into account (i) water needs for humans, (ii) environmental water requirements, and (iii) the basin's ability to assimilate pollution [34]. On the other hand, we suggest the study of forest-climate-water interactions as a hydraulic analogy since changes in forest cover can alter precipitation at regional scales [25].

2. A water-forest interface through hydraulic analogy

To address water-forest restoration perspectives, it is worth studying the water-forest system. It can be shown as a transpiration system which is analogously analyzed as any other closed system that transports fluids, for example, a pipeline. **Figure 1** indicates, from a hydraulic similarity, the energy grade line concept applied to liquid and vapor phases of the system. Because of low velocities, velocity head is neglected due to laminar flow. Energy is required to (1) extract water, (2) transport water through the soil, (3) transport it through the plant, (4) vaporize water, and (5) transport water into the atmosphere. All these energy phases have losses. In **Figure 1**, the change in the point-to-point energy level gives the head losses and energy sources. Flow direction is toward the negative energy gradient. Accounting for the head losses through the liquid phase of the transpiration process, the total frictional head loss is.

$$h_f = h_s + h_r + h_x + h_l, \tag{1}$$

in which h is the dissipated head with the subscripts "s," "r," and "x" and "l" refers to the soil, root, xylem, and the leaf components, respectively. These subscripts will designate the same flow components when used hereafter with other hydraulic terms. Assuming that the flow is laminar, Darcy's flow (qDarcy = k·h/L) and continuity equation (Q = qDarcy·a) may be applied to each component of flow in the liquid phase. For any given hydraulic component, *qDarcy* is the velocity, *k* is the hydraulic conductivity, *L* is the length toward the flow directs, *a* is the cross-sectional area to the flow, and *Q* is the total flow rate. If the expression h = Q/a · L/k is substituted into each component, it yields the hydraulic factors that affect transpiration in the liquid phase.

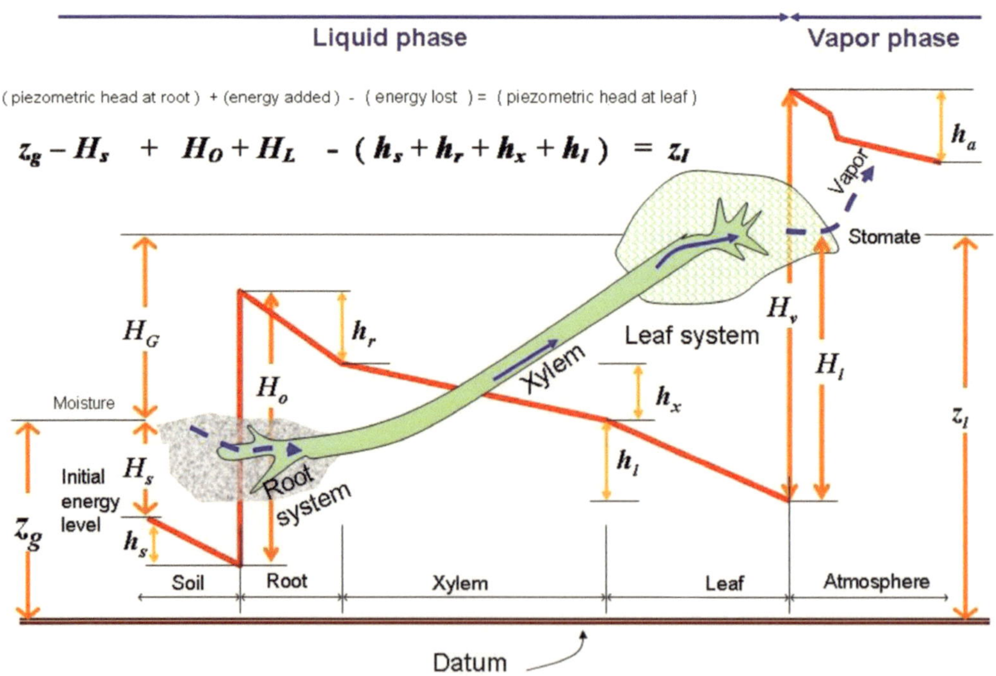

Figure 1. Hypothetical energy grade line in the transpiration process for condition $h_f = H_f$ (adapted from [35]; Ross and Salisbury [36]; and Stewart et al. [37]). See explanation in the text.

$$h_f = Q \cdot \left[\frac{L_s}{a_s k_s} + \frac{L_r}{a_r k_r} + \frac{L_x}{a_x k_x} + \frac{L_l}{a_l k_l} \right] \tag{2}$$

The terms within brackets in Eq. (2) are resistance terms and could be replaced as *Rs* for the soil's resistance and *Rp* for the plant's resistance, as follows:

$$R_s = \frac{L_s}{a_s k_s} \tag{3}$$

$$R_p = \frac{L_r}{a_r k_r} + \frac{L_x}{a_x k_x} + \frac{L_l}{a_l k_l} \tag{4}$$

in which *Rs* and *Rp* are the equivalent hydraulic resistances of the soil and plant, respectively. *Rs* is dependent on the soil moisture content and the type of soil, and *Rp* depends on the type of the plant and the stage of growth that includes the extent of the root system development. From Eq. (2), a more simplified equation is obtained:

Figure 2. The 12 major river basins in Brazil. Source: [38].

$$Q = \frac{h_f}{R_s + R_p}, \tag{5}$$

which states that the flow rate that is delivered in the liquid phase is dependent on the ratio of the total head available and the hydraulic resistances of the soil (depending on water content) and the plant (depending on the vegetation type and growth phase). Eq. (5) shows that the lower the value for *Rp* and the more head, *hf*, that can be developed, the more drought resistant the plant is.

It should be mentioned that the vertical distance between the plant-conduit and the energy grade line (*EGL*) represents the pressure head because the velocity head is negligible. Furthermore, the water-forest cycle shown in **Figure 2** is the key element for restoration and conservation measures, such as BMPs, in different spatiotemporal scales and under several scenarios of climate and land-use changes explained in the following sections.

3. Water resource conservation strategies for South American watersheds: Examples from Brazilian watersheds

Brazil is the largest South American country and is the fifth largest country in the world (both geographically and in population). It presents 87% of urban population and most of the population lives near the Atlantic coast in the east [39]. Due to Brazil's large area (8.5 million km^2), each region presents different meteorological patterns and biomes. Moreover, several river basins were selected for hydropower production, representing ca. 87% of all the energy demand in Brazil [40].

3.1. Hydrometeorological aspects of Brazilian watersheds

The hydrological and meteorological characteristics of the major Brazilian river basins, some of them for hydropower generation, are presented in **Table 1** and **Figures 2 and 3**.

A mixture of climate zones characterizes Brazil [41]. These climate zones can be divided into (1) atmospheric circulation; (2) thermic regions, which are related to the monthly extreme temperatures; and (3) categories related to droughts (**Figure 3**).

The Amazon river basin (3,870,000 km^2) occupies around 45% of the Brazilian territory. The mean flow in the region corresponds to 74% of the national flow in Brazil (179.516 m^3/s). In spite of its abundance of water, just a minor part of the Brazilian population lives in this region, with a demographic density of 2.51 hab./km^2 (10 times less than the national mean), and the water demand is very reduced, as it is only 3% of the national water demand [38].

The Paraná river basin (879,873 km^2), which occupies 10% of the Brazilian territory and includes the metropolitan regions of Sao Paulo and Curitiba, represents the most economically developed region in Brazil and has a high population density, approximately 69.7 hab./km^2. It presents the highest water resource demand in the country, near 31% of the national demand.

River basin	Drainage area (km^2)	Mean precipitation (mm/year)	Mean discharge (m^3/s)	Specific discharge (L/s/km^2)	Evapotranspiration (mm/year)	Native vegetation cover (%)
Major Brazilian river basin regions						
Amazon	3,870,000	2205	132,145	34.1	1128	85
East Atlantic	388,160	1018	1484	3.8	897	35
Western Northeast Atlantic	274,300	1700	2608	9.5	1400	48
Eastern Northeast Atlantic	286,800	1052	774	2.7	967	50
Southeast Atlantic	214,629	1401	3167	14.8	936	31
Southern Atlantic	187,552	1644	4055	21.6	962	39
Paraguay	363,446	1359	2359	6.5	1154	58
Parana	879,873	1543	11,831	13.4	1119	16
Parnaiba	333,056	1064	767	2.3	991	75
São Francisco	638,466	1003	2846	4	862	53
Tocantins-Araguaia	920,000	1774	13,779	15	1302	53
Uruguay	274,300	1623	4103	15	1151	31
Selected strategic river basins for hydropower generation in Southeast Brazil						
Cantareira	2279	1475	35	15	988	26
Emborcação	29,076	1485	456	16	991	29
Três Marias	51,576	1404	657	13	1003	20
Furnas	52,197	1484	905	17	937	19
Mascarenhas	71,649	1238	890	12	847	19

Table 1. Hydrometeorological aspects of the major Brazilian river basin regions and selected watersheds draining to strategic reservoirs for hydropower in Southeastern Brazil.

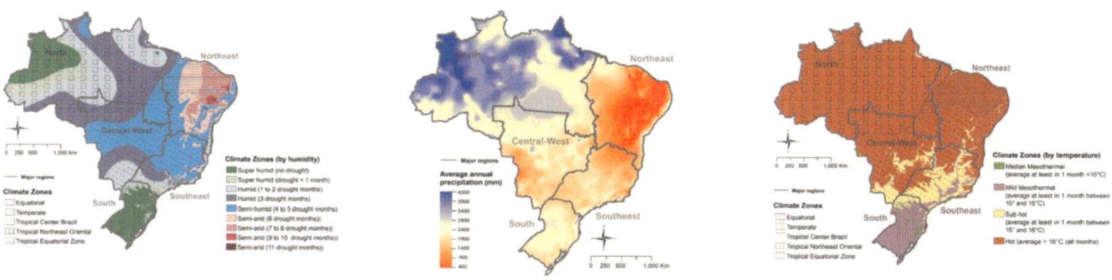

Figure 3. The complex blend of climate zones by average annual precipitation, humidity, and temperature in Brazil. Source: Bressiani et al. [42]. Reproduced with the permission of the authors.

However, the mean flow in the region represents only 6.6% of the national flow [38]. This fact reveals the imbalance in the distribution of the water resources in the country. The Amazon region also presents the higher pluviometric index of the country, 25% higher than the national mean. This occurs because the Amazon river basin is located in a hot and humid climate region, classified as equatorial (**Figure 3**). This region is characterized by the presence of equatorial air masses, of the continental type, by the action of the intertropical convergence zone (ICZ), formed by the convergence of trade winds [43].

In contrast, the *Atlântico Leste* river basin (388,160 km^2), the *Atlântico Nordeste Oriental* (286,800 km^2), Parnaíba (333,056 km^2), and Sao Francisco (638,466 km^2), present minor annual mean precipitations, as a consequence of hot and dry climates, with sparse and irregular rains and a mean rainfall of 500 mm per year [43].

Regarding vegetation, most of the Amazon river basin is covered by its native vegetation, consisting of the Amazon (approximately 87% of its original cover), the "Cerrado," or Savannah (approximately 60% of the original vegetation). On the other hand, the Paraná is the basin which presents the smallest area covered by native vegetation proportionally. In comparison to the original area, only 18% of Savannah and 15% Atlantic Forest biomes remain. The Uruguay (274,300 km^2) and *Atlântico Sudeste* river basins also drastically reduced their native vegetal cover, as shown in **Table 1**.

Table 1 also provides data of the drainage area of important reservoirs for the Southeast region of Brazil. The Cantareira Water Supply System (2300 km^2), hereafter referred to as the Cantareira System, encompasses 1000 hm^3 of reservoirs and is the main source of water supply for the metropolitan region of Sao Paulo and Campinas [38]. This region (**Figures 4** and **5**) was severely affected by the water crisis in 2013–2015, which brought water supply problems to the metropolitan region of Sao Paulo and Campinas and hydroelectric power generation concerns throughout the country [44–48].

Regarding **Table 1**, drainage areas of reservoirs Emborcação, Três Marias, Furnas, and Mascarenhas, which are essential for hydroelectric power generation, irrigation, and water supply, have less than 30% of native forest cover [40]. The Emborcaçao reservoir is the most affected with high deforestation rates (only 6% of the original cover remains). On the other hand, the Cantareira System, considered with the lowest deforestation rate, only has 33% of native Atlantic Forest, partially due to watershed restoration programs (see [29]).

3.2. Biomes in Brazil

Brazil is the country with the highest biodiversity of vegetation in the world. There are more than 55,000 cataloged plant species of an estimated total ranging between 350,000 and 550,000 [49]. A significant part of this biodiversity is found in the Atlantic Forest, a biome that stands out for its high levels of richness, endemism, and devastation. Despite having 20,000 species of vascular plants [50], of which between 7000 and 8000 are endemic [51], only 11% of the Brazilian Atlantic Forest still remains [52]. Since Brazil's colonization, the Atlantic Forest deforestation has narrowed the delivery of ecosystem services. This progressive devastation

http://dx.doi.org/10.5772/intechopen.82526

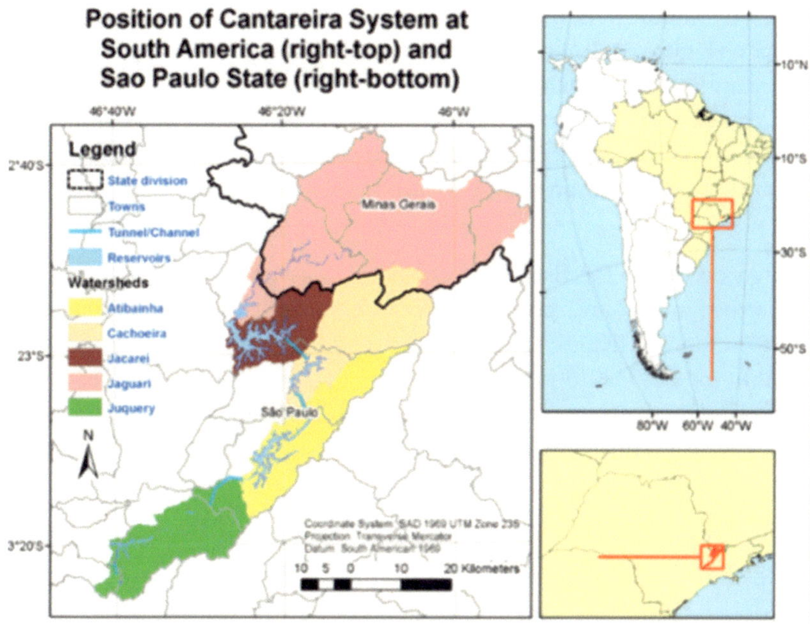

Figure 4. Watersheds of the Cantareira water supply system (left) and its position in South America (top right) and in Sao Paulo state (bottom right).

Figure 5. A general view of the Cantareira system region, a 2300-km^2 drainage area connected to the 1000 hm^3 of reservoirs in the anthropized Atlantic Forest biome. Photo by Denise Taffarello, December 2012.

was caused by the exploration of forest resources, advancements of agricultural borders and by coastal urbanization, as well as a zone going between 40 and 50 km into the inland [53].

The large area of Brazil encompasses six biomes. They consist of the following Brazilian names: *Amazônia, Caatinga, Cerrado, Mata Atlântica, Pampa,* and *Pantanal*.

First, the Amazon rainforest is the largest forest in the world, conditioned by the humid equatorial climate (**Figure 3**). It represents around 35% of the forest areas globally. However, recent predatory agricultural practices, new acts, and decrees have (1) reduced environmental licensing requirements, (2) suspended the ratification of indigenous lands, (3) reduced the size of protected areas, and (4) allowed land grabbers to obtain the charters of deforested areas [54]. This has led Amazon deforestation to 17%, which makes it difficult for Brazil to fulfill the Paris Agreement.

Second, Caatinga occurs in the Brazilian Northeast. Its vegetation is formed by palm trees which usually grow in dry and poor (in terms of nutrients) soils. Rossato et al. [55] used weather data from the CPTEC/INPE platform to estimate the Palmer Drought Severity Index (PDSI) in Brazil in the 2000–2015 period and found that the PDSI achieved severe to extreme dry scales over time in the Northeast, where the dry conditions are a socioeconomic and environmental problem. They concluded that the PDSI is useful to assess different soil moisture water conditions and design risk maps.

Third, the Cerrado, which presents diverse regions, ranging from clean fields devoid of woody vegetation to *cerradão*, a dense tree formation, is also in danger [54].

Fourth, the Atlantic rainforest encompasses 35% of Brazilian's biodiversity and boasts high levels of species richness but also has critical rates of deforestation. Only 11–16% of the Brazilian Atlantic Forest still remains on the coastline [51], and the hydrometeorological patterns of the region are very different. However, the presence of humid winds from the ocean is remarkable, and it favors vegetation development.

Fifth, the Pampa is composed of different herbaceous species, and in some areas, their environment is integrated with several *Araucaria* trees, in the South region of Brazil.

Last but not least, the Pantanal is an alluvial plain influenced by rivers that drain the Upper Paraguay basin, where it develops a fauna and flora of rare beauty and abundance. The flood regimes are seasonal, and during the increased flows of the Paraguay river, the water chemistry changes depending on the mineral composition of parent material, soil use, and vegetation cover. Consequently, not only flow direction and magnitude fluxes change but also transparency, temperature, and the macro-ionic composition of the water. These environmental variabilities induce a habitat pattern that influences the composition of aquatic communities. It favors those that have adopted strategies to exist within a specified range of environmental conditions. Karr and Chu [56] described five dynamic environmental factors that regulate the structure and functioning of any aquatic ecosystems shown in **Figure 6**. These factors can be applied to explaining, for example, the beauty and high abundance of organisms found in the Pantanal biome.

3.3. Some Latin America projects for watershed restoration

Research has shown that degraded watersheds are related to higher poverty levels [57]. Thus, restoration and creation of wetlands have been recommended for the development of human populations in an integrated way [58]. Most restoration projects have multiple purposes regarding the quality of the biological community and the hydrological functioning of the system.

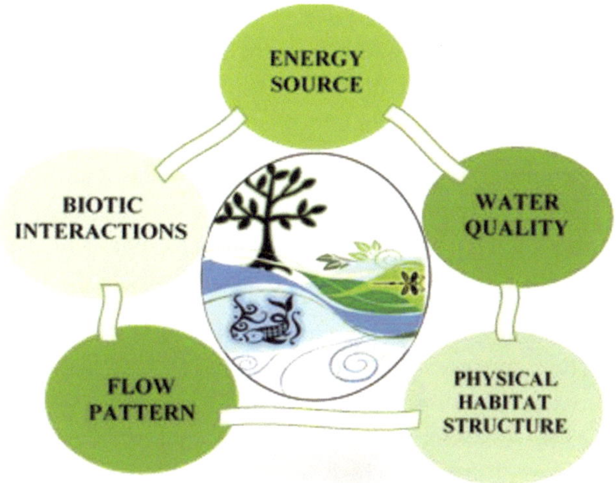

Figure 6. The dynamic environmental components structuring ecosystem functioning. Source: Adapted from Karr and Chu [56].

However, the nature of ownership in forested areas (e.g., whether private or government property) greatly affects forest management and, consequently, affects the water yield, water quality, and their services. In Latin America, forest plantation, that is, *Eucalyptus* and *Pinus*, mostly occurs on private land, whereas native forests prevail in public areas. Moreover, the Brazilian government has recently signed decrees, which potentially threat native forests, due to political bargaining [54]. This fact is very worrisome to the risks associated with hydrological extremes, climate/environmental changes, and losses of ecosystem services.

The most common actions to enhance hydrological services and resilience front disturbances, which can be conducted in both private and public lands, are hydrological control, wetland construction, denitrification barriers, biogeochemical barriers, and food web manipulation. A successful restoration strategy, according to the Society for Ecological Restoration (2011), consists of the steps highlighted below.

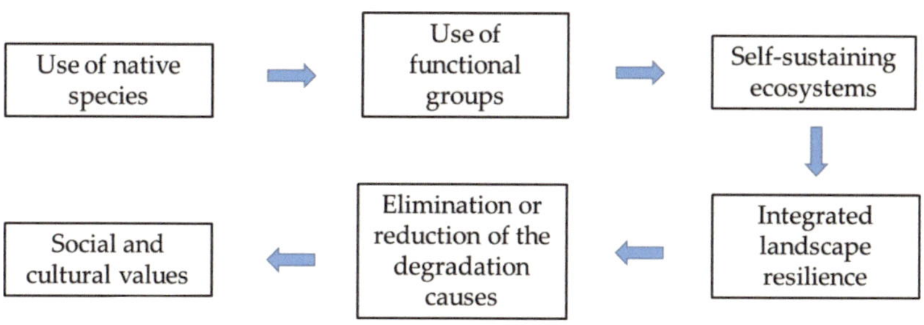

Thus, there are some initiatives in Latin America which can achieve this goal (the watershed restoration), mainly through EbA methods.

Latin American countries were pioneers in development and implementation of PES projects as a kind of EbA method. There is a wide range of initiatives, which apply PES on various scales, in different contexts and with diverse specific objectives [59], serving as a general mechanism to align economic investments in human and ecosystem welfare [60].

Previous formal PES programs began in Cauca Valley, Colombia, in the mid-1990s, where silvopastoral practices were used in a pilot project to protect upper watersheds [61, 62]. PES research reflects the watershed's particularities, different topographic, soil, and climate conditions, that hold a major part of the world's unique biodiversity [63].

Costa Rica was the first country to establish a formal PES, called *Programa de Pagos por Servicios Ambientales*, in 1997. There are governmental subsidies to help water users (such as hydropower companies) to pay land owners for benefits generated by their conservation actions in the watersheds. In Quito (Ecuador), the water and hydropower companies pay for the protected area's upstream withdrawal, which is a source of a significant amount of clean water [64]. Currently, there are EbA projects under ongoing implementation in all the 20 countries of Latin America [63, 65]; however the freshwater PES experiences that have been developed still act on a very small scale, contributing to the conservation and restoration of a small area related to the total area in each country).

The first publication about ecosystem services in Latin America is from 1997: a researcher from the National Institute for Research in the Amazon (INPA) proposed a strategy for achieving sustainable development in rural Brazilian Amazonia, which required both short-term and long-term measures [66]. Later, the improved strategy was proposed for the Cocibolca Lake watershed, Nicaragua, showing four scenarios built on the *Soil and Water Assessment Tool* (SWAT) model to reduce the potential of sediments and nutrient loads. Currently, various initiatives have spread throughout Latin America, especially the platform called "Water Funds" [60, 65], but the majority are still small scale and present failures in the forest growth and hydrological monitoring.

3.3.1. Some Brazilian examples

The Brazilian Atlantic Forest is a biodiversity hotspot in the world and constitutes a carbon sink. For these reasons, it offers an economical opportunity for establishing restoration or conservation practice [67, 68]. One of these initiatives is the Atlantic Forest Restoration Pact [69, 70], a public-private partnership with the aim to restore 150,000 km^2 of forest by 2050 using native species. Another initiative was the "Produtor de Água/PCJ Project" [71]. This project aims to stimulate actions of forest restoration, conservation of fragments, and soil conservation practices on private properties to provide remuneration to the farmers to create and/or maintain ecosystem services [72].

The following EbA projects are or were implemented in the five regions of Brazil:

- "Bolsa Floresta Program" (North)
- "Monte Pascoal-Pau Brasil Project" (Northeast)

- "Manancial Vivo Program" (Central West)
- "Mina d'Água Project" (Southeast)
- "Corredores Ecológicos Chapecó-Timbó Project" (South), to quote only a few examples

There are public-private partnerships working with EbA for the restoration of watersheds and carbon sink, as well as private companies and nongovernmental organization initiatives. In some of these, there is support from Brazilian universities in the ecohydrological monitoring of the projects [46, 47]. See the EbA initiatives in the Brazilian Atlantic Forest developed until 2015. You can find more details on these and more initiatives in the paper by Taffarello et al. [29] (**Table 2**).

Starting year	Number of cases	Project's total area (km^2)	Investment value (million US $)	Average payment value per project (US$/ha/year)	Adaptive measures considered for long-term changes
1997	1	0.5	1.4	77–254	No
2005	1	28.5 (by 2011)	2.986	77.5 UFEX (March/2010)	Yes. Monitoring partnership with Esalq-USP, EESC-USP, IAG-USP, and Viçosa Federal University Fiscal Unit of Extrema municipality (UFEX in Portuguese) (1 UFEX = USD 0.70 in 2018)
2006	3	20.9; 1640; 14.8	1.112	33–242	Partially. Vegetation, hydrographic, and property management monitoring occur in the Oasis Project
2008	1	32.3	2.846	88	No. The *Bolsa Verde* operations manual (IEF, 2010) does not mention monitoring
2009	8	12.48; 36.77; 1.12; 22.22; 19.62; 10.00; 0.41; 31.99	14.828	4.4–968	Yes. The Camboriu Project and Water Producer/PCJ (EESC-USP), through water quantity (rainfall and runoff measures); water quality (turbidity, total suspended solids, pH, dissolved oxygen, electrical conductivity, organic matter, and nutrients such as ammonia, nitrogen, nitrates, and total phosphates); hydrologic health (geomorphologic analyses of the water bodies' structure)
2010	4	0.39; 11.6; 11,000; 8.8	3768	17.79–633.43	Partially. Mina d'Água: monitoring plan and impact assessment developed with the support of specialists from the World Bank
2012	1	Priority areas of the town for public supply		25 UFM 1 UFM = US$ 46,62	No

Source: Adapted from [29].

Table 2. Case studies of EbA projects for watershed restoration at the Brazilian Atlantic Forest.

Current research has been developed to improve existing methodologies and market-based policy tools to identify the generation and maintenance of the ecosystem services by the watershed restoration [73, 74].

The potential provision of the ecosystem hydrological services depends on the equilibrium of the hydric balance, namely, the relation between the hydric availability and demand (variable given natural oscillations or induced by impacts from anthropic activities), besides the state and functional distribution of the ecosystems on the watersheds. From this interaction, the composition "water+climate" is the principal element of sustainability [75, 76], directly influencing the biodiversity. Therefore, the ecosystemic approach is a strategy for the integrated management of soil, water, and biodiversity, promoting a balanced conservation and sustainable use of natural resources.

3.3.2. Some Colombian examples

The Paramo biome is a set of neotropical alpine grassland ecosystems covering the upper region of the northern Andes. It plays a key role in the hydrology [77–79]. It is characterized by elevations between 3000 and 5000 miles above sea level (MASL) and a constant mean monthly temperature with large diurnal temperature fluctuations. The precipitation patterns in the Paramo are exceedingly complex in terms of amount and seasonality; the precipitation varies from approximately 600 to 4400 mm from a bimodal pattern to a unimodal one depending on the location [80]. Over the last years, these ecosystems have been strongly impacted by human interventions and climate change [77, 81–85]. This has shown its

Water fund name	Phase, stage	Year founded	Ecosystem description	Conservation target area (ha)
Bogotá, Water We are	Operation	2009	Paramo system and high Andean forests	60,000
Valle del Cauca, Water for Life and Sustainability Foundation	Operation	2009	Some Paramo areas, high Andean forests, and inter-Andean valleys	65,000
Medellín, Green Basin	Operation	2013	High Andean forests and Paramo	22,300
Cali, Madre Agua	Creation	2015	Humid forest, tropical forest, cloud forest, and Paramo	4550
Cúcuta, Biocuenca Alliance	Creation	2015	Paramo system and high Andean forests	15,900
Cartagena Water Fund	Creation	2016	Riparian wetland	138
Sierra Nevada de Santa Marta	Idea and prefeasibility	TBD	Mountainous coastal system isolated from the Andes and Paramo	TBD
Santa Marta y Ciénaga	Idea and prefeasibility	TBD		TBD
Santander, Bucaramanga	Idea and prefeasibility	TBD	Paramo system and high Andean forests	TBD

Table 3. Colombian water funds description projects (modified from [86]).

vulnerability and importance as a water supply source of some of the main cities in South America. Therefore, initiatives, such as PES, are tools that in recent years have been implemented in Colombia, but they are not yet common practice. Recently the Colombian government, through law 870 of May 2017, established the payment for environmental services and other incentives for conservation. Among the most recognized PSE implementation projects in Colombia are the Water Funds (see **Table 3**), an initiative led by The Natural Conservancy (TNC), which is currently benefiting nearly 12,295,247 million people [86]. These Water Funds projects, which are in the operational phase, were established to benefit populations from the Valle del Cauca (Water for Life and Sustainability Water Fund Cauca Valley, Southwestern Colombia) and the cities of Medellin (*Cuenca Verde*) and Bogotá (*Agua Somos*), all these with the particularity of being developed for a complex and fragile Paramo Andean ecosystem [80]. While in the creation and feasibility phase, there are funds such as Cucuta (Biocuenca Alliance), the Cartagena Water Fund aiming to conserve the riparian wetland areas and the Sierra Nevada de Santa Marta Fund, a mountainous coastal system isolated from the Andes, and Santander, designed for water conservation in the metropolitan area of Bucaramanga with major conflicts over mining exploitation in the Paramo area [87, 88].

4. Conclusions

Relationships between water and forests depend on the soil characteristics, including moisture dynamics, which in turn impacts the water security and overall sustainability of water resource management. As a result of the interaction among soil, water cycles, forest, and climate, we address further parameters and guidance for the conservation of the hydrologic and forest resources in the watersheds. In this chapter, we discussed ecohydrological processes and the associated ecosystem services provided by the catchments and promising opportunities for watershed restoration. In this context, we argue that EbA strategies can help to develop the economy of Latin American countries, where the population is expected to increase more and more over the next few years. Such strategies would require the creation or expansion of markets for ecosystem services, hydrologic and forest participative monitoring (e.g., through hydrosociology and citizen science), human resource development, and training. Thus, linking water and vegetation is essential to secure diverse hydrometeorological services and the resilience of the biodiversity hotspots. In South America's biomes, these services can be used to optimize annual costs and benefits of conservation and provide financial support for restoration projects in most affected communities. Not only South American society's demands, but also environmental needs in Latin America in general, can be achieved through holistic and transdisciplinary PES projects, some of them briefly summarized in this chapter. The PES initiatives can also increase income and, to a certain extent, boost employment rates and community development. For example, the Water Funds can help comprehend the relationships between water yield and forests through "research-for-action" initiatives. This integrated management can reduce people's and ecosystems′ vulnerability, as well as increase their resilience to cope with global change impacts.

Acknowledgements

We are grateful to the CAPES for the postdoctoral scholarship to the first author (CAPES/PNPD) and CAPES PROEX (PPG-SHS, EESC/USP). Also, this work was supported by the National Institute of Science and Technology for Climate Change Phase 2 under CNPq Grant 465501/2014-1, the FAPESP Grant 2014/50848-9, and the National Coordination for High Level Education and Training (CAPES) Grant 16/2014. We would like to thank the civil engineer, Dario J. P. Macedo, for some information on Brazilian watersheds and Marina Bittar (scientific initiation undergraduate) for some help. Some South American research-for-action initiatives can be found at INCT-MC2-FAPESP, INCLINE/USP, and CEPID/CeMEAI-FAPESP projects, at the official sites, respectively, of: http://www.bv.fapesp.br/pt/auxilios/97629/inct-2014-inct-para-mudancas-climaticas-inct-mc/; http://www.incline.iag.usp.br/data/index_USA.php; and http://www.bv.fapesp.br/en/auxilios/58570/cemeai-center-for-mathematical-sciences-applied-to-industry/.

Conflict of interest

The authors declare that they have no conflict of interest.

Author details

Denise Taffarello[1]*, Diego Alejandro Guzman Arias[2], Danielle de Almeida Bressiani[3,4], Davi Gasparini Fernandes Cunha[1], Maria do Carmo Calijuri[1] and Eduardo Mario Mendiondo[1]

*Address all correspondence to: taffarellod@gmail.com

1 Department of Hydraulics and Sanitation, São Carlos School of Engineering, University of Sao Paulo, Sao Carlos (SP), Brazil

2 Department of Civil Engineering, Pontificia Bolivariana University, Bucaramanga, Colombia

3 Brazilian Meteorological Agency Ltda. (Climatempo), São Paulo, Brazil

4 Federal University of Pelotas (UFPel), Pelotas, Brazil

References

[1] Food and Agriculture Organization of the United Nations (FAO). OCDE-FAO Perspectivas Agricolas 2016-2025. OCDE-FAO Perspectivas Agrícolas. Paris: OECD Publishing; 2016. DOI: 10.1787/agr_outlook-2016-es. [Accessed: September 30, 2018]

[2] Crutzen PJ. Geology of mankind. Nature. 2002;**415**(6867):23-23. DOI: 10.1038/415023a

[3] Waters CN, Zalasiewicz J, Summerhayes C, Barnosky AD, et al. The Anthropocene is functionally and stratigraphically distinct from the Holocene. Science. 2016;**351**(6269), aad2622. DOI: 10.1126/science.aad2622

[4] Magurran AE. How ecosystems change. Science. 2016;**351**(6272):448-449

[5] Ehret U, Gupta HV, Sivapalan M, et al. Advancing catchment hydrology to deal with predictions under change. Hydrology and Earth System Sciences. 2014;**18**:649-671

[6] Brauman KA. Hydrologic ecosystem services: Linking ecohydrologic processes to human well-being in water research and watershed management. Wires-Water. 2015;**2**: 345-358

[7] Brauman KA, Daily GC, Duarte TK, Mooney HA. The nature and value of ecosystem services: An overview highlighting hydrologic services. Annual Review of Environment and Resources. 2007;**32**:67-98. DOI: 10.1146/annurev.energy.32.031306.102758

[8] Dodds WK, Perkin JS, Gerken JE. Human impact on freshwater ecosystem services: A global perspective. Environmental Science & Technology. 2013;**47**:9061-9068. DOI: 10.1021/es4021052

[9] Postel SL, Daily GC, Ehrlich PR. Human appropriation of renewable fresh water. Science. 1996;**271**:785-788

[10] Vörösmarty CJ, Mcintyre PB, Gessner MO, Dudgeon D, Prusevich A, Green P, et al. Global threats to human water security and river biodiversity. Nature. 2010;**467**:555-561

[11] Tansley A. The use and abuse of vegetational concepts and terms. Ecology. 1935;**16**(3): 284-307. Retrieved from http://www.jstor.org/stable/1930070 [Accessed: May 6, 2016]

[12] Jørgensen SE. Ecohydrology as an important concept and tool in environmental management. Ecohydrology & Hydrobiology. 2016;**16**(1):4-6

[13] Olden JD, Kennard MJ, Pusey BJ. A framework for hydrologic classification with a review of methodologies and applications in ecohydrology. Ecohydrology. 2012;**5**(4):503-518

[14] Rodriguez-Iturbe I. Ecohydrology: A hydrologic perspective of climate-soil-vegetation dynamics. Water Resources Research. 2000;**31**(1):3-9

[15] Zalewski M, Robarts RD. Ecohydrology—A new paradigm for integrated water resources management. SIL News. 2003;**40**:1-5

[16] Kobiyama M, Genz F, Mendiondo EM. Geo-bio-hidrologia. In: Kobiyama M, Genz F, Mendiondo EM, editors. I Fórum de Geo-Bio-Hidrologia: Estudo em Vertentes e Microbacias Hidrográficas. Brazil: UFPR; 1998. pp. 1-25

[17] Daily GC. Introduction: What Are Ecosystem Services? In: Nature's Services - Societal Dependence on Natural Ecosystems. Washington: Island Press; 1997. pp. 1-10

[18] MILLENIUM ECOSYSTEM ASSESSMENT. Ecosystems and Human Well-Being: Synthesis. Washington, DC: Island; 2005

[19] Hannah D, Sadler JP, Wood PJ. Hydroecology and ecohydrology: A potential route forward? Hydrological Processes. 2007;**21**(24):3385-3390

[20] Janauer GA. Ecohydrology: Fusing concepts and scales. Ecological Engineering. 2000;**16**(1): 9-16

[21] Zalewski M. Ecohydrology—The scientific background to use ecosystem properties as management tools toward sustainability of water resources. Ecological Engineering. 2000; **16**(1):1-8

[22] Zalewski M, Wagner I. Integrated Watershed Management—Manual of Ecohydrology and Phytotechnology. Osaka, Japan: UNEP-IHP, United Nations Environment Programme-International Hydrologic Program; 2004. 479 p

[23] Mendiondo EM. Challenging issues of urban biodiversity related to ecohydrology. Brazilian Journal of Biology. 2008;**68**(4):983-1002. DOI: 10.1590/S1519-69842008000500007

[24] Cardinale BJ, Duffy JE, Gonzalez A, Hooper DU, Perrings C, Venail P, et al. Biodiversity loss and its impact on humanity. Nature. 2012;**486**(7401):59-67

[25] Gordon LJ, Peterson GD, Bennett EM. Agricultural modifications of hydrological flows create ecological surprises. Trends in Ecology & Evolution. 2008;**23**(4):211-219

[26] Ricklefs RE. A Economia da Natureza. 6th ed. Rio de Janeiro: Guanabara Koogan; 2015. Translation of the Economy of Nature. Translator: Pedro P. de Lima-e-Silva, Translation coordinator: Cecília Bueno

[27] CBD—Convention on Biological Diversity: X/33 biodiversity and Climate Change, Decision Adopted by the Conference of the Parties to the Convention on Biological Diversity at its Tenth Meeting; UNEP/CBD/COP/DEC/x/33; 29 October 2010; Nagoya, Japan: Secretariat of Convention on Biological Diversity; 2010

[28] Creed IF, van Noordwijk M. Forest and Water on a Changing Planet: Vulnerability, Adaptation and Governance Opportunities. [S.l.]: International Union of Forest Research Organizations (IUFRO), 2018. 192 p. Available at: https://www.iufro.org/es/publications/series/world-series/article/2018/07/10/world-series-vol-38-forest-and-water-on-a-changing-planet-vulnerability-adaptation-and-governan/ [Accessed: August 21, 2018]

[29] Taffarello D, Calijuri MC, Viani RAG, Marengo JA, Mendiondo EM. Hydrological services in the Atlantic Forest, Brazil: An ecosystem-based adaptation using ecohydrological monitoring. Climate Services. 2017;**2**:1-16

[30] BFN/GIS Federal Agency for Nature Conservation/DEUTSCHE GESELLSCHAFT FÜR INTERNATIONALE ZUSAMMENARBEIT (GIZ) GMBH. Natural solutions to climate change: The ABC of ecosystem-based adaptation. Summary and Conclusions from an International Expert Workshop held 4–9 August 2013 on the Isle of Vilm, Germany; 2013

[31] Taffarello D, Srinivasan R, Mohor GS, Guimarães JLB, Calijuri MDC, Mendiondo EM. Modelling freshwater quality scenarios with ecosystem-based adaptation in the

headwaters of the Cantareira system, Brazil. Hydrology and Earth System Sciences. 2018; **22**:4699-4723. DOI: 10.5194/hess-22-4699-2018

[32] Tundisi JG. Recursos hídricos no futuro: Problemas e soluções. Estudos Avançados. 2008; **22**(63):7-16. http://www.scielo.br/pdf/ea/v22n63/v22n63a02.pdf

[33] Hoekstra AY, Chapagain AK, Aldaya MM, Mekonnen MM. The Water Footprint Assessment Manual: Setting the Global Standard. London, UK: Earthscan; 2011

[34] Mekonnen MM, Hoekstra AY. Global gray water footprint and water pollution levels related to anthropogenic nitrogen loads to fresh water. Environmental Science & Technology. 2015;**49**:12860-12868

[35] Hendricks D, Hansen VE. Mechanics of evapo-transpiration. Journal of the Irrigation and Drainage Division: Proceedings of the American Society of Civil Engineering. 1962;**88**(1R2): 67-82

[36] Ross C, Salisbury F. Plant physiology, 1947–1972. Annals of the Missouri Botanical Garden. 1974;**61**(1):112-131. DOI: 10.2307/2395187, 25 Years of Botany (1974)

[37] Stewart DW, Dwyer LM, Desjarnis RL. A mathematical model of transpiration using nonlinear least square analysis. Canadian Agricultural Engineering. 1985;**27**(1):1-6

[38] National Water Agency. Brazilian Water Resources Report [Relatório de Conjuntura dos Recursos Hídricos do Brasil]; Brasília: Agência Nacional de Águas (ANA); 2015

[39] CIA. Central Intelligence Agency United States of America. 2018. Available from: https://www.cia.gov/library/publications/resources/the-world-factbook/geos/print_br.html [Accessed: July 20, 2018]

[40] Zhang R, Cuartas LA, de Castro Carvalho LV, Leal KRD, Mendiondo EM, Abe N, et al. Season-based rainfall–runoff modelling using the probability-distributed model (PDM) for large basins in southeastern Brazil. Hydrological Processes. 2018;**32**:2217-2230. DOI: 10.1002/hyp.13154

[41] Nimer E. Climatologia Do Brasil [Climatology of Brazil]. 2nd ed. Rio de Janeiro: IBGE, Departamento de Recursos Naturais e Estudos Ambientais; 1989. 421 p

[42] Bressiani DA, Gassman PW, Fernandes JG, Garbossa LHP, Srinivasan R, Bonumá NB, et al. Review of soil and water assessment tool (SWAT) applications in Brazil: Challenges and prospects. International Journal of Agricultural and Biological Engineering. 2015;**8**(3): 9-35. DOI: 10.3965/j.ijabe.20150803.1765

[43] Torres FTP, Machado PJO. Introdução à Climatologia [Introduction to Climatology]. São Paulo, Brazil: Cengage Learning; 2011. p. 256

[44] Guzman DA, Mohor GS, Taffarello D, Mendiondo EM. Economic impacts of drought risks for water utilities through severity-duration-frequency framework under climate change scenarios. Hydrology and Earth System Sciences Discussions. 2017:1-39. DOI: 10.5194/hess-2017-615

[45] Nobre CA, Marengo JA, Seluchi ME, Cuartas LA, Alves LM. Some characteristics and impacts of the drought and water crisis in Southeastern Brazil during 2014 and 2015. Journal of Water Resource and Protection. 2016;**08**(02):252-262. DOI: 10.4236/jwarp.2016.82022

[46] Taffarello D, Guimarães J, Lombardi R, Calijuri MC, Mendiondo EM. Hydrologic monitoring plan of the Brazilian water producer/PCJ project. Journal of Environmental Protection. 2016;**7**:1956-1970. DOI: 10.4236/jep.2016.712152

[47] Taffarello D, Mohor SG, Calijuri MC, Mendiondo EM. Field investigations of the 2013–14 drought: Quali-quantitative freshwater monitoring at the headwaters of Cantareira system, Brazil. Water International. 2016;**41**(5):776-800. DOI: 10.1080/02508060.2016.1188352

[48] Coutinho RM, Kraenkel RA, Prado PI. Catastrophic regime shift in water reservoirs and São Paulo water supply crisis. PLoS One. 2015;**10**(9):1-14. DOI: 10.1371/journal.pone.0138278

[49] Simões CMO, Schenkel EP, Gosmann G, Mello JCP, Mentz LA, Petrovick PR. Farmacognosia: Da Planta Ao Medicamento [Pharmacognosy: From Plant to the Drug], UFSC Editor; Porto Alegre, Brazil: UFRS, 2007

[50] Myers N, Mittermeier RA, Mittermeier CG, da Fonseca GAB, Kent J. Biodiversity hotspots for conservation priorities. Nature. 2000;**403**:853-858. DOI: 10.1038/35002501

[51] Viani RAG, Benini R, Padovezi A, Veiga-Neto F. Mecanismos de pagamentos por serviços ambientais Para a restauração Florestal da Mata Atlântica [payment for ecosystem services mechanisms for forest restoration in the Atlantic Forest] In: Ministério Público de São Paulo, Editor. Temas de Direito Urbanístico 6 - Áreas de Risco. São Paulo: Imprensa Oficial; 2012. p. 357-382

[52] Ribeiro MC, Metzger JP, Martensen AC, Ponzoni FJ, Hirota MM. The Brazilian Atlantic Forest: How much is left, and how is the remaining forest distributed? Implications for conservation. Biological Conservation. 2009;**142**:1141-1153. DOI: 10.1016/j.biocon.2009.02.021

[53] Ab'sáber AN. Os Domínios de Natureza no Brasil: Potencialidades Paisagísticas [The Nature's Domains in Brazil: Landscape Potentialities]. Sao Paulo: ateliê Editorial; 2003. 158 p

[54] Rochedo PRR, Soares-Filho B, Schaeffer R, Viola E, Szklo A, Lucena AFP, et al. The threat of political bargaining to climate mitigation in Brazil. Nature Climate Change. 2018;**8**:695-698. DOI: 10.1038/s41558-018-0213-y

[55] Rossato L, Marengo JA, de Angelis CF, Pires LBM, Mendiondo EM. Impact of soil moisture over palmer drought severity index and its future projections in Brazil. Brazilian Journal of Water Resources. 2017;**22**:e36. DOI: 10.1590/2318-0331.0117160045

[56] Karr JR, Chu EW. Sustaining living rivers. Hydrobiologia. 2000;**422**:1-14. DOI: 10.1023/A:1017097611303

[57] Kumar JLG, Zhao YQ. A review on numerous modelling approaches for effective, economical and ecological treatment wetlands. Journal of Environmental Management. 2011;**92**:400-406. DOI: 10.1016/j.jenvman.2010.11.012

[58] Comín FA, Romero JA, Hernández O, Menéndez M. Restoration of wetlands from abandoned rice fields for nutrient removal, and biological community and landscape diversity. Restoration Ecology. 2001;**9**:201-208

[59] Pagiola S, von Glehn HC, Taffarello D, Editors. Experiências de Pagamentos Por Serviços Ambientais no Brasil. São Paulo: Secretaria do Meio Ambiente; 2013. 338 p

[60] Goldman-Benner R, Benítez S, Boucher T, Calvache A, Daily G, Kareiva P, et al. Water funds and payments for ecosystem services: Practice learns from theory and theory can learn from practice. Oryx. 2012;**46**:55-63. DOI: 10.1017/S0030605311001050

[61] Echavarría M. Financing watershed conservation: The FONAG water fund in Quito, Ecuador. In: Pagiola S, Bishop J, Landell-Mills N, editors. Selling Forest Environmental Services: Market-Based Mechanisms for Conservation and Development. 10th ed. London: Earthscan; 2002

[62] Pagiola S, Zhang W, Colom A. Can payments for watershed services help finance biodiversity conservation? A spatial analysis of Highland Guatemala. Journal of Natural Resources Policy Research. 2010;**2**(1):7-24. DOI: 10.1080/19390450903350812

[63] Balvanera P et al. Ecosystem services research in Latin America: The state of the art. Ecosystem Service: Science, Practice and Policy. 2012;**2**:56-70. DOI: 10.1016/j.ecoser.2012.09.006

[64] Southgate D, Wunder S. Paying for watershed Services in Latin America: A review of current initiatives. Journal of Sustainable Forestry. 2009;**28**:497-524. DOI: 10.1080/10549810902794493

[65] Bremer LL, Auerbach DA, Goldstein JH, Vogl AL, Shemie D, Kroeger T, et al. One size does not fit all: Natural infrastructure investments within the Latin American water funds partnership. Ecosystem Services. 2016;**17**:217-236. DOI: 10.1016/j.ecoser.2015.12.006

[66] Fearnside PM. 1997. Environmental services as a strategy for sustainable development in rural. Amazonia. Ecological Economics. 1997;**20**(1):53-70. DOI: 10.1016/S0921-8009(96)00066-3

[67] Joly CA, Rodrigues RR, Metzger JP, Haddad CFB, Verdade LM, Oliveira MC, et al. Biodiversity conservation research, training, and policy in São Paulo. Science. 2010;**328**:1358-1359. DOI: 10.1126/science.1188639

[68] Brancalion PHS, Viani RAG, Strassburg BBN, Rodrigues RR. Finding the money for tropical forest restoration. Unasylva. 2012;**63**(239):41-50

[69] Rodrigues RR, Lima RAF, Gandolfi S, Nave AG. On the restoration of high diversity forests: 30 years of experience in the Brazilian Atlantic Forest. Biological Conservation. 2009;**142**:1242-1251. DOI: 10.1016/j.biocon.2008.12.008

[70] Melo FPL, Pinto SRR, Brancalion PHS, Castro PS, Rodrigues RR, Aronson J, Tabarelli M. Priority setting for scaling-up tropical forest restoration projects: Early lessons from the Atlantic Forest restoration pact. Environmental Science & Policy, 2013;**33**:395-404. DOI: 10.1016/j.envsci.2013.07.013

[71] Padovezi A, Viani RAG, Kubota U, Taffarello D, Faria M, Bracale H, et al. Produtor de água na bacia hidrográfica Piracicaba/Capivari/Jundiaí. In: Pagiola S, von Glehn HC, Taffarello D, editors. Experiências de Pagamentos Por Serviços Ambientais no Brasil. Ed. São Paulo: Secretaria do Meio Ambiente; 2013. pp. 99-113

[72] Mazzocato C, Taffarello D, Mendiondo EM. Ecohidrologia Para monitoramento do pagamento por serviços ambientais no estado de São Paulo [Ecohydrology for the monitoring of payment for ecosystem services in Sao Paulo state]. In: Simposio Brasileiro de Recursos Hídricos [Brazilian Symposium of Water Resources]; 22-25 November 2013; Bento Gonãalves, Brazil, 2013

[73] Taffarello D, Mendiondo EM. Adaptation Options from Ecohydrology and Water Footprint to the Payment for Ecosystem Services in the Context of River Restoration Projects in Brazil. Em: 4th Conference on Ecological Restoration, Mérida. Society for Ecological Restoration; 2011

[74] Muradian R, Rival L. Between markets and hierarchies: The challenge of governing ecosystem services. Ecosystem Services. 2012;**1**:93-100. DOI: 10.1016/j.ecoser.2012.07.009

[75] Moss B, Battarbee RW, Kernan M. Changing climate and changing planet. In: Moss B, Battarbee RW, Kernan M, editors. Climate Change Impacts on Freshwater Ecosystems. Oxford: Wiley Blackwell; 2010. pp. 1-15

[76] Tucci CEM, Mendes CA. Curso de Avaliação Ambiental Integrada de Bacia. Ministério do Meio Ambiente—Secretaria de Qualidade Ambiental—Rhama Consultoria Ambiental. Porto Alegre, Brazil: Rhama Environmental Consultant. 311 p. 2006

[77] Buytaert W, Célleri R, De Bièvre B, Cisneros F, Wyseure G, Deckers J, et al. Human impact on the hydrology of the Andean páramos. Earth-Science Reviews. 2006;**79**:53-72. DOI: 10.1016/j.earscirev.2006.06.002

[78] Cárdenas MF, Tobón C. Recuperación del funcionamiento hidrológico de ecosistemas de páramo en Colombia. Revista U.D.C.A Actualidad & Divulgación Científica. 2017;**20**:403-412

[79] Cárdenas MF, Tobón C, Buytaert W. Contribution of occult precipitation to the water balance of páramo ecosystems in the Colombian Andes. Hydrological Processes. 2017;**31**: 4440-4449. DOI: 10.1002/hyp.11374

[80] Arroyo M, Cavieres L. High-elevation Andean ecosystems. In: Encyclopedia of Biodiversity. 2nd ed. 2013. pp. 96-110

[81] Anderson EP, Marengo J, Villalba R, Halloy S, Young B, Cordero D, et al. Consequences of climate change for ecosystems and ecosystem services in the tropical Andes. Climate Change and Biodiversity in the Tropical Andes. 2011;**1**:1-18. DOI: 10.13140/2.1.3718.4969

[82] Balthazar V, Vanacker V, Molina A, Lambin EF. Impacts of forest cover change on ecosystem services in high Andean mountains. Ecological Indicators. 2015;**48**:63-75. DOI: 10.10 16/j.ecolind.2014.07.043

[83] Buytaert W, Cuesta-Camacho F, Tobón C. Potential impacts of climate change on the environmental services of humid tropical alpine regions. Global Ecology and Biogeography. 2011;**20**:19-33. DOI: 10.1111/j.1466-8238.2010.00585.x

[84] Herzog SK, Martínez R, Jørgensen PM, Tiessen H, editors. Climate Change and Biodiversity in the Tropical Andes. 2011;**1**:1-348. DOI: 10.13140/2.1.3718.4969

[85] Mora DE, Campozano L, Cisneros F, Wyseure G, Willems P. Climate changes of hydrometeorological and hydrological extremes in the Paute basin, Ecuadorean Andes. Hydrology and Earth System Sciences. 2014;**18**:631-648. DOI: 10.5194/hess-18-631-2014

[86] Goldman-Benner RL, Benitez S, Calvache A, Ramos A, Veiga F. Water funds: A new ecosystem service and biodiversity conservation strategy. In: Encyclopedia of Biodiversity. Elsevier Ltd., 2013;352-366. http://dx. doi.org/10.1016/B978-0-12-384719-5.00330-0

[87] Betancur-Corredor B, Loaiza-Usuga JC, Denich M, Borgemeister C. Gold mining as a potential driver of development in Colombia: Challenges and opportunities. Journal of Cleaner Production. 2018;**199**:538-553. DOI: 10.1016/j.jclepro.2018.07.142

[88] Pérez-escobar OA, Cámara-leret R, Bateman R, Earle SA, Wright DJ, Safina C, et al. Mining threatens Colombian ecosystems. Science Letters. 2018;**359**:6383

Chapter 5

Geo-Statistical Assessment of the Intensity, Duration, Frequency and Trend of Drought over Gangetic West Bengal, Eastern India

Krishna Gopal Ghosh

Additional information is available at the end of the chapter

http://dx.doi.org/10.5772/intechopen.80037

Abstract

This chapter presents spatio-temporal (1901–2002) appraisal of the intensity, duration, frequency and trend of drought over Gangetic West Bengal (GWB), Eastern India using standardized precipitation index (SPI). The study reveals that, after 1950s the magnitude of deficit precipitation has increased substantially. Stepping up of the mean intensity of most intense drought events; average drought duration; severe and extreme drought frequency in this agricultural tract at the latter half of the twentieth century are also some alarming events. The western degraded plateau is more sensitive to extreme droughts but, the impact is expected to be rigorous over the adjacent areas. In an nutshell this work provides the evidences demonstrating the intensification of aridity in the northern Rarh plain and moribund delta which may corresponds to degradation and lowering of water resources especially ground water which may also lead to increase of socio-economic vulnerability to drought. Such altered hydrolo-meteorological system hence calls for review of the agricultural practices and water use in this counterpart.

Keywords: standardized precipitation index (SPI), drought intensity, drought duration, drought frequency, threshold rainfall

1. Introduction

Monsoon region of South Asia remains one of the important worries with respect to frequency and magnitude of drought in the contemporary scenario of climate change [1]. About 23 million hectares of Asian rice producing areas experience frequent yield loss due to drought [2]. Afghanistan, India, Pakistan and Sri Lanka have reported droughts at least once in every

3 year over the past 5 decades [3]. In India the net sown area is about 140 Mha, out of which as much as 50% area is considered as severely drought prone [4]. Therefore, a comprehensive assessment of drought is needful for monsoon based agro-economy of India.

Gangetic West Bengal (GWB), the leading agricultural hubs of Eastern India, severely experienced the effect of climate change over the last few years [5]. Late monsoon arrival has been observed with less intensity, duration of summer has become longer and drought has become more frequent [6–10]. Moreover, this region is less experienced of coping with droughts resulting in poor preparedness. Growing population, lacking water resource management initiatives etc. further compounded the problem. Therefore, we need to improve our knowledge on drought jeopardy in this densely populated tract with vast agricultural expanse. The present chapter is an attempt in this regard to ensure two folds objectives—i. to portray a comprehensive and holistic picture of droughts over GWB-its intensity-duration-frequency and trend and ii. to identify areas exposed to drought.

2. Geographical personality of the study area

The study focuses on the southern half of West Bengal below Farakka barrage (**Figure 1**) located between latitudes 21°32′23″N to 24°51′20″N and longitudes 85°49′49″ E to 89°8′48″ E (area: 63,879 km^2, elevation range: 0–677 m) surrounded by Jharkhand in the West, Odisha in the Southwest and Bangladesh in the East. Physiographically, GWB forms the transitional zones between Chhotanagpur plateau in the West and Ganga-Brahmaputra delta in the southern and eastern section. River Bhagirathi and its tributaries/distributaries drain this region [11]. The climate is typical sub-tropical monsoon type having four main seasons namely, winter (Jan-Feb); Pre-Monsoon (Mar-May); Monsoon (Jun-Sep) and Post-Monsoon (Oct-Dec) [12]. Out of the total annual rainfall, about 70–80% occurs during the monsoon and contributing as much as 90% to the discharge of the rivers.

GWB is a densely populated (1051 person/km^{-2}) tract, coupled with vast stretch of fertile alluvial soil and is the heart of rice and jute cultivation as well as freshwater fish production of eastern India, the gross cropped area and cropping intensity of this region count about 112 and 184% respectively [13]. Agriculture in this region is mainly rain-fed and rainfall extremities put heavy stress on not merely agricultural activities but also other economic activities. Average water demand in this region varies from about 0.9 to 1.8mm^3 per km^{-2} in compare to the average water availability of 0.5 to 1.0mm^3 per km^{-2} [14]. These demonstrate sensitiveness to drought of this region.

According to the study of Ghosh [9] there is a considerable decrease in rainfall during early monsoonal month June and mid monsoonal month August in this tract. In the year 2010 10 districts of GWB have received <33% of the normal monsoon rainfall, which severely affected the sowing of paddy [15]. In the northern Rarh and moribund delta there are significant decreasing trend of rainfall (**Figure 2**) which is a sign of strengthening of drier condition of these two regions. Nath et al. [16], WBSAPCC [6], RPAPCC [8] etc. have roughly addressed

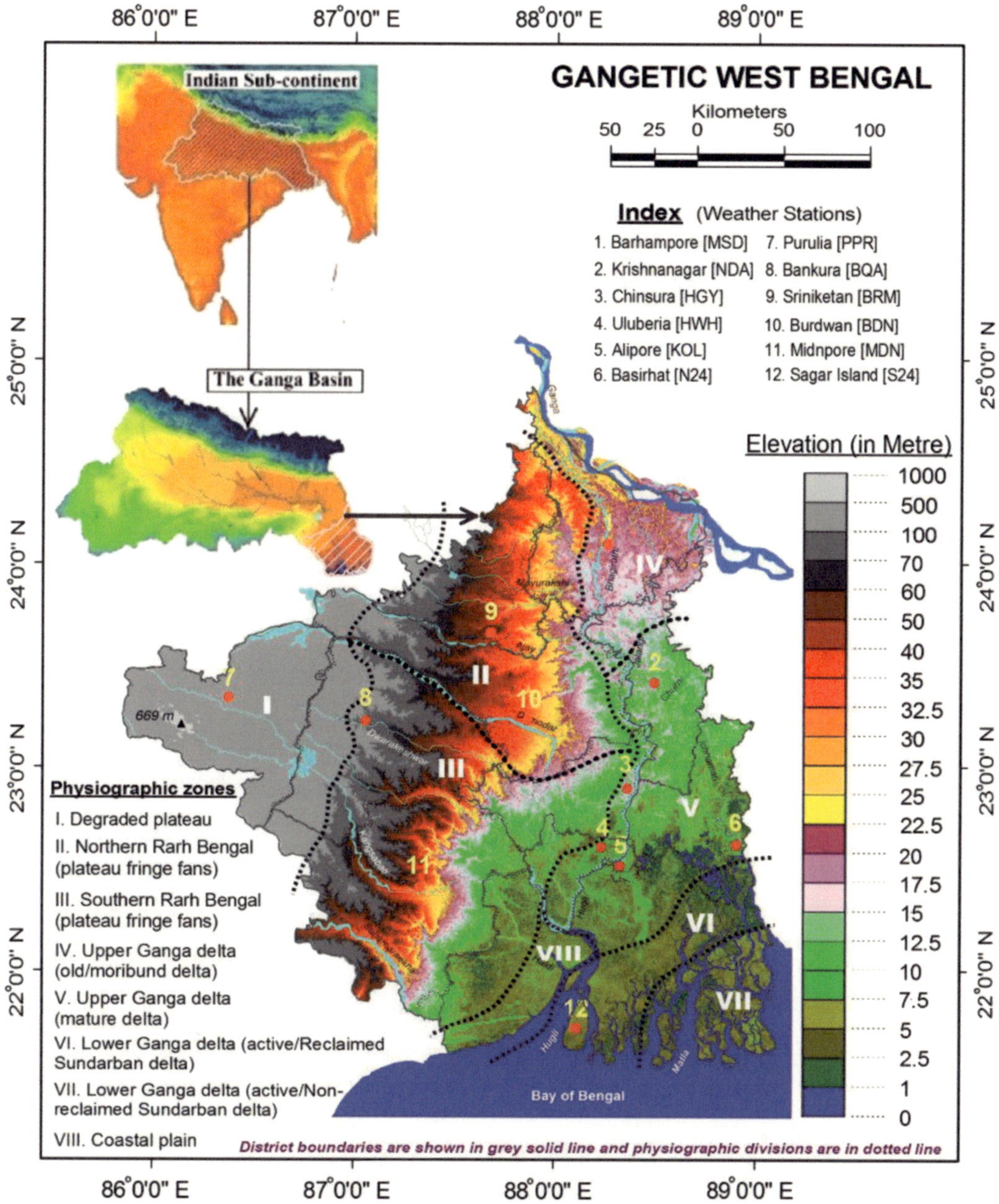

Figure 1. Reference map of the Gangetic West Bengal showing elevations, district boundaries, physiographic divisions and location of the weather stations used in the analysis.

the way for drought management of the state. Nevertheless, comprehensive assessment of drought jeopardy of this region in prime prerequisite before chalking out the management plan and the present chapter opt to do so.

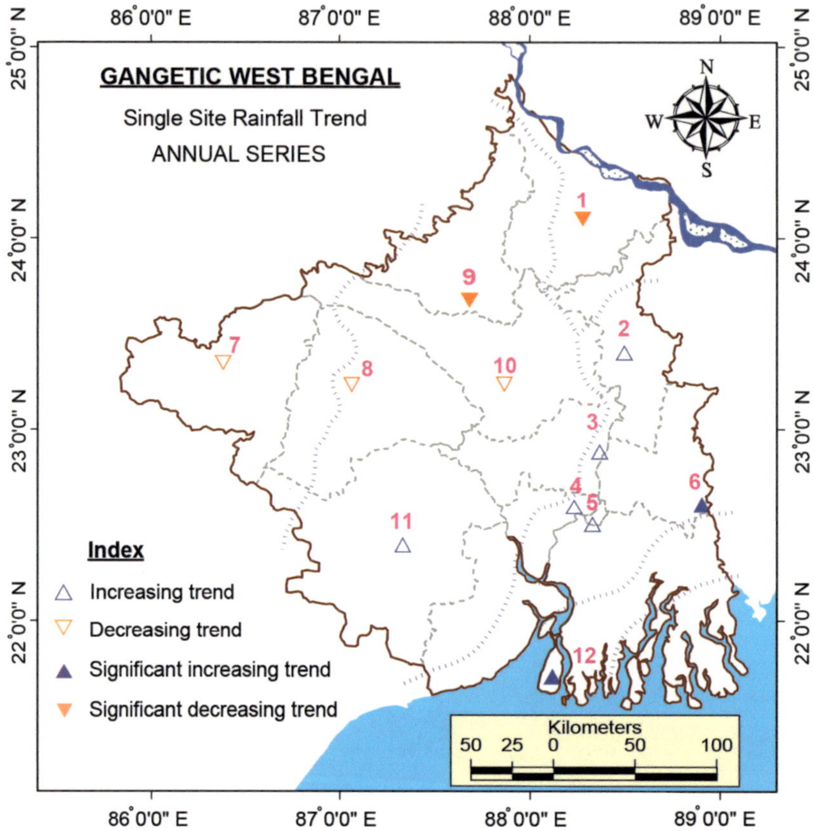

Figure 2. Site (meteorological station) specific annual rainfall trend during 1901–2002.

3. Materials and methods

For the present study continuous time series rainfall data [http://www.indiawaterportal.org/] belonging to 12 meteorological stations (one representative station per district) of the GWB (**Figure 1**) for the period of 1901–2002 have been used. The Indian Meteorological Department (IMD) assembles the data as well as controls the data quality.

Over a geographic area deficit of rainfall from normal during a period is broadly accepted as drought. In this study standardized precipitation index (SPI) has been used to detail the geographical variations of drought at multiple time steps. Different dimensions of drought (like- drought intensity, duration, frequency etc.) following Dracup et al. [17] have been captured for the holistic appraisal.

3.1. Standardized precipitation index (SPI)

SPI, developed by McKee et al. [18, 19] is a simple but flexible tool to monitor drought at multiple time step. SPI is recommended by the WMO as a standard drought monitoring index

[20] and is calculated by taking the difference of the precipitation, Xi from the mean, $\bar{X}$ for a particular time step, and then dividing it by the standard deviation, σ [21].

$$SPI = \frac{Xi - \overline{X}}{\sigma} \quad (1)$$

SPI is used to monitor both dry and wet conditions [22]. Negative values indicate dry and positive values indicate wet periods [23]. As SPI becomes more negative or positive, the conditions become more severely dry or wet (**Table 1**).

McKee et al. [18] originally calculated the SPI for 3-, 6-, 12-, 24- and 48-month timescales. For the present case, typically SPI for 3- and 12-months step SPI are calculated to explore the drought variation at inter-seasonal and inter-annual time scales.

3.2. Drought evaluation parameters

Drought intensity (I_D): I_D annotates departure of a climate index from its normal value [24]. According to McKee et al. [18] a drought event is defined as a period in which the SPI is continuously negative and SPI reaches a value of −1.0 or less. Hence, I_D indicates the absolute value of SPI less than −1.0. Lesser the value more will be the drought intensity.

Drought duration (D_D): D_D equals the number of months between its start and end [25]. A drought event starts when the SPI is continuously negative and reaches an intensity of −1.0 or less while, the event ends when the SPI becomes positive.

Drought magnitude (M_D) and mean intensity, (MI_D): M_D corresponds to the cumulative water deficit over a drought period [26] and the average of this cumulative water deficit over the drought period is MI_D. Thus, M_D is the absolute value of the sum of all SPI values during a drought event and MI_D refers to magnitude divided by duration.

Drought frequency (F_D): F_D is used to assess the drought liability during a study period [27]. The number of droughts per 100 years was calculated as:

SPI values	Draught severity class	D-scale
2.0+	Extremely wet	W3
1.5 to 1.99	Very wet	W2
1.0 to 1.49	Moderately wet	W1
−.99 to .99	Near normal	N
−1.0 to −1.49	Moderately dry	D1
−1.5 to −1.99	Severely dry	D2
−2 and less	Extremely dry	D3

Source: http://www.wamis.org/agm/pubs/SPI/WMO_1090_EN.pdf.

Table 1. Precipitation excess (wet) or deficit (dry) severity class according to SPI values.

$$F_{D\,i,100} = \frac{Ni}{i.n} \times 100\ (\%) \tag{2}$$

where $F_{Di,100}$ is the frequency of droughts for timescale i in 100 years; Ni is the number of months with droughts for timescale i in the n-year set; i is timescale (3-, 6-, 12-, 24-months); n is the number of years in the data set.

Trend analysis: The rank-based nonparametric Mann-Kendall test [28–29] is applied to the long-term data in this study to detect statistically significant trends. Sen's nonparametric method [30] is used to estimate the trends slope in the time series data.

Return period or recurrence interval (Tr): Bonaccorso et al. [31] expressed Tr as a function of the statistical characteristics of historical long records of precipitation and of a threshold parameter. In the present study the original concept of the return period [32] is used, i.e. the average number of years between events above a threshold magnitude.

Rainfall threshold/critical rainfall (T_{RD}): To calculate the threshold rainfall, first we have to calculate the $\bar{x}$ and σ (Eq. (1)) for a particular time step (say, for 3 month Dec–Jan–Feb of 30 years of continuous rainfall data), for a given station and specify the desired SPI value (say −1.0) in Eq. (3) for which critical rainfall is to be calculated.

$$Xi = \sigma SPI + \overline{X} \tag{3}$$

3.3. Data processing and related calculations and mapping

The complete set of raw data for the said period (Jan 1901 to Dec 2002) in the current study have been tested to check if there are any missing data or irregularities in the data series. SPI is then calculated for different time steps using 'SPI Calculator' of the National Drought Mitigation Centre (NDMC) as recommended by WMO [20]. Afterwards drought-related indicators as indicated in Section 6.2.2 are calculated. SPSS 14.0 and XLSTAT 2015 Excel plug-in have been used for the MK test and Sen's slope. For the purpose of revealing the spatial variation over GWB Choropleth maps are then prepared using GIS software.

4. Results and discussion

4.1. Spatial and temporal assessment of the drought intensity

4.1.1. Time series assessment of draught intensity

On 12-month time lag 10 significant droughts have occurred during 1901–2002 (**Figure 3a**) among which, 1966–1967 is the most significant with D_D of 14 months; PI_D of −2.82 and a MI_D of −2.06. On 48-month time step the last century roughly exhibit some consecutive surplus and deficit phase (**Figure 3b**): (i) slight deficit (1901–1917), (ii) short surplus (1918–1922); (iii) oscillating or near normal (1923–1940); (iv) short surplus (1941–1953); (v) longest and peak deficit (1954–1970); (vi) peak surplus (1971–1982); (vii) short deficit (1983–1986) and (viii) longest surplus (1987–2001). Noticeably, at the longer time scales, droughts become less frequent but their duration increases (**Figure 3b**) and after 1950s the extremities of surplus and deficit as well as duration have increased substantially. Most of the drought event reaches to its maximum intensity during the pre-monsoon months of March to May (**Table 2**).

4.1.2. Spatial character of the drought intensity

During the 102 year time span the MI_D at the 3- and 12-month time step was −0.81 and − 0.77 respectively. MI_D for most stations (69.44%) at all the time scale were ≥−0.8 but did not cross

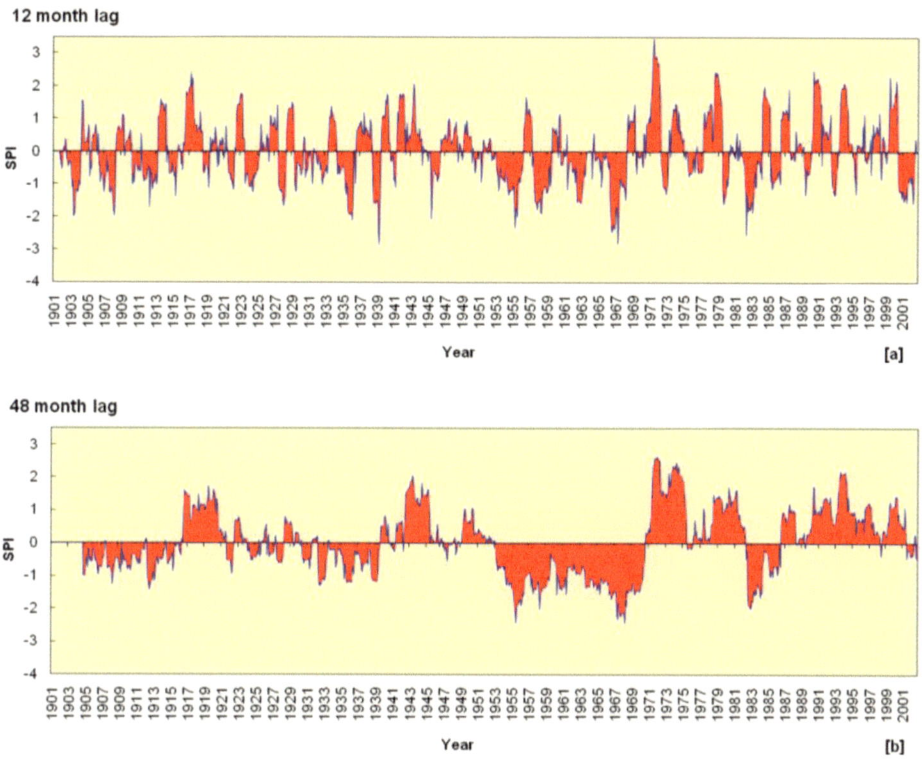

Figure 3. SPI time series (1901-2002) for GWB for (a) 12- and (b) 48- month time scales.

Lag time	Observed peak intensity (PI_D) during 1901–2002			Mean intensity (MI_D)
	SPI	Year	Month	
3-Month	−3.72	1999	April	−2.68 (gross avg. −0.81)
6-Month	−3.65	1922	March	−2.64 (gross avg. −0.77)
12-Month	−2.83	1939	June	−1.68 (gross avg. −0.77)
24-Month	−2.62	1968	January	−2.13 (gross avg. −0.81)
48-Month	−2.43	1968	May	−1.49 (gross avg. −0.79)

Table 2. Drought intensity at different time scale for GWB during 1901–2002.

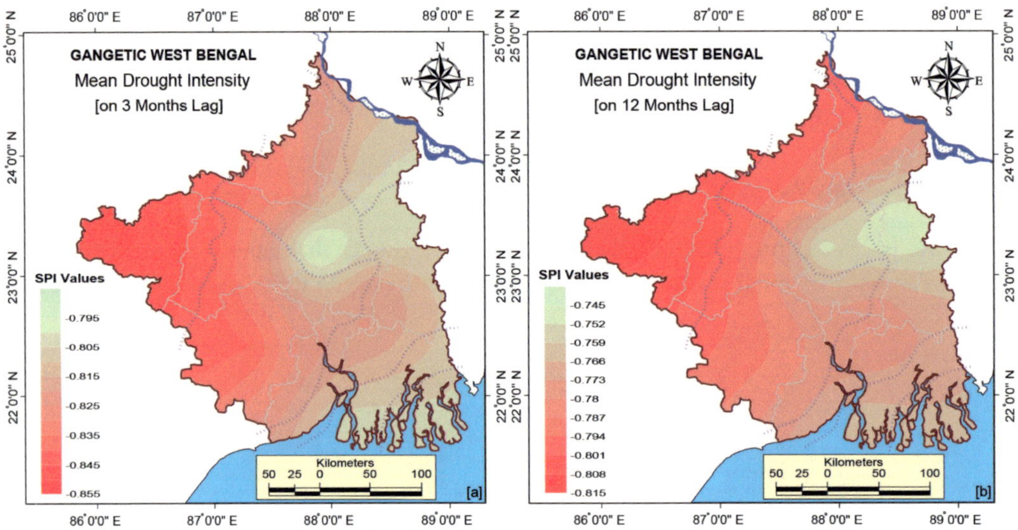

Figure 4. Iso-mean intensity drought map for (a) 3-, (b) 12- month SPI.

the limit of mild drought (i.e. 0 to −0.99). Regional MI_D varies between −0.75 (Krishnanagar) and −0.86 (Purulia). Spatially (**Figure 4a** and **b**), average drought intensity is the greatest in the degraded plateau region and western fringe of the Rarh region; it is less in the mature & active delta region and eastern fringe of the Rarh plain. Roughly the drought intensity gradually decreases from West to East and South East.

At the 3-month step observed peak intensities are maximum in the mature delta region (**Figure 5a** and **b**). At the 12-month step it was greatest in the western degraded plateau region, extreme northern part of moribund delta and Rarh plain region. Thus the western degraded plateau region is more sensitive to long-duration droughts.

4.2. Duration and magnitude of drought events

During 1901–2002, average drought duration identified by SPI on 12-month scale for most stations (>90.0%) was 4–6 months (**Table 3**) and the regional average value was 4.18. The maximum drought duration at all the stations was more than 20 months (**Table 3**) and the regional average of the most intense drought was 7 month. The average magnitude of the longest drought and the most intense drought event on 12 month scale is about −29 and −16 respectively with their mean intensity of −1.37 and −2.29.

Spatially at the shorter time span (3-month lag) the coastal plain followed by the southern Rarh and parts of the lower Ganga plain are sensitive to relatively longer drought duration (**Figure 6a**). At the 12-month lag, it was relatively longer in the degraded plateau and plateau fringe fans of the Rarh Bengal (**Figure 6b**). Noticeably the area suffered from lengthier drought, magnitude was also counted high there (**Figure 7**).

http://dx.doi.org/10.5772/intechopen.80037

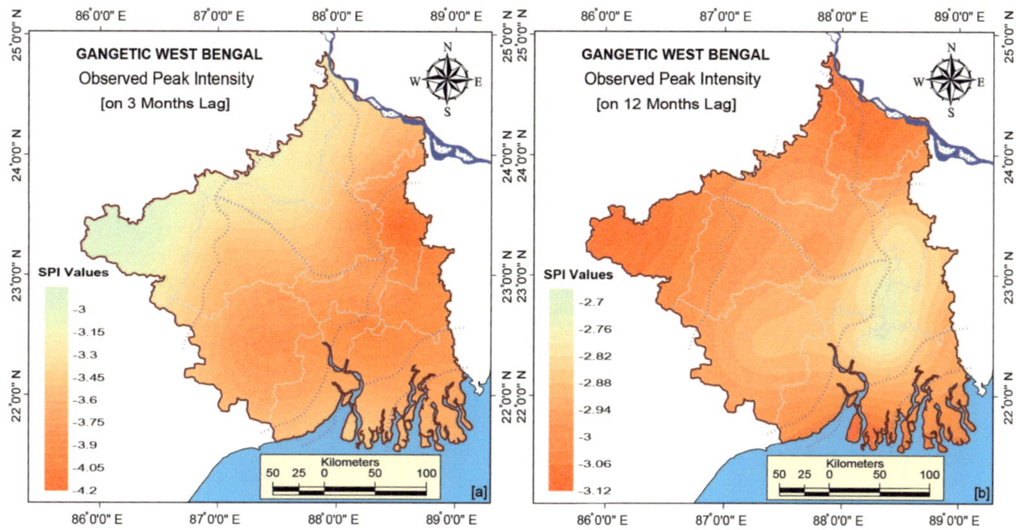

Figure 5. Observed maximum intensity drought map at (a) 3-month and (b) 12-month time scale.

Weather station	Longest duration (≤ − 1.0 for consecutive months)			Most intense duration (≤ − 2.0 for consecutive months)			$D_DA(M)$
	$D_DL(M)$	Year	M_D	$D_DL(M)$	Year	M_D	
Berhampore	24	Sept, 2000 to Aug, 2002	−32.18	11	July, 1982 to May, 1983	−27.87	4.57
Krishnanagar	23	July, 1966 to May, 1968	−42.33	8	Sept, 1982 to April, 1983	−18.09	4.19
Chinsurah	21	Sept, 1957 to May, 1959	−28.13	7	Oct, 1935 to April, 1936	−15.53	3.93
Uluberia	20	Oct, 1957 to May, 1959	26.05	8	Oct, 1935 to May, 1936	−19.97	4.46
Alipore	21	Sept, 1957 to May, 1959	−28.28	7	Oct, 1935 to April, 1936	−16.72	3.98
Basirhat	23	Sept, 1957 to July, 1959	−37.63	8	Oct, 1935 to May, 1936	−21.02	4.55
Purulia	25	Aug, 2000 to Aug, 2002	−46.23	12	July, 1966 to June, 1967	−33.34	5.17
Bankura	23	Sept, 2000 to July, 2002	−36.58	12	July, 1966 to June, 1967	−31.59	5.08
Sriniketan	23	Sept, 2000 to July, 2002	−34.06	7	Aug, 1966 to Feb, 1967	−16.15	5.15
Burdwan	23	Sept, 2000 to July, 2002	−31.73	8	Aug, 1966 to Mar, 1967	−19.4	4.55
Midnapore	20	Sept, 1957 to April, 1959	−25.75	8	Sept, 1935 to April, 1936	−20.55	5.07

Weather station	Longest duration (≤ − 1.0 for consecutive months)			Most intense duration (≤ − 2.0 for consecutive months)			$D_DA(M)$
	$D_DL(M)$	Year	M_D	$D_DL(M)$	Year	M_D	
Sagar Island	20	Oct, 1957 to May, 1959	−24.52	9	Sept, 1935 to May, 1936	−29.43	4.17
GWB	21	Sept, 1957 to May, 1959	−28.76	7	Sept, 1966 to Mar, 1967	−16.05	4.18

Note: ***DD(M):*** duration (month); ***DDL(M):*** observed longest duration (month); ***DDI(M):*** observed most intense duration (month); ***DDA(M):*** average duration (month); ***MD:*** magnitude.

Table 3. Drought duration and magnitude at 12 month time scale for different weather stations of the GWB during 1901–2002.

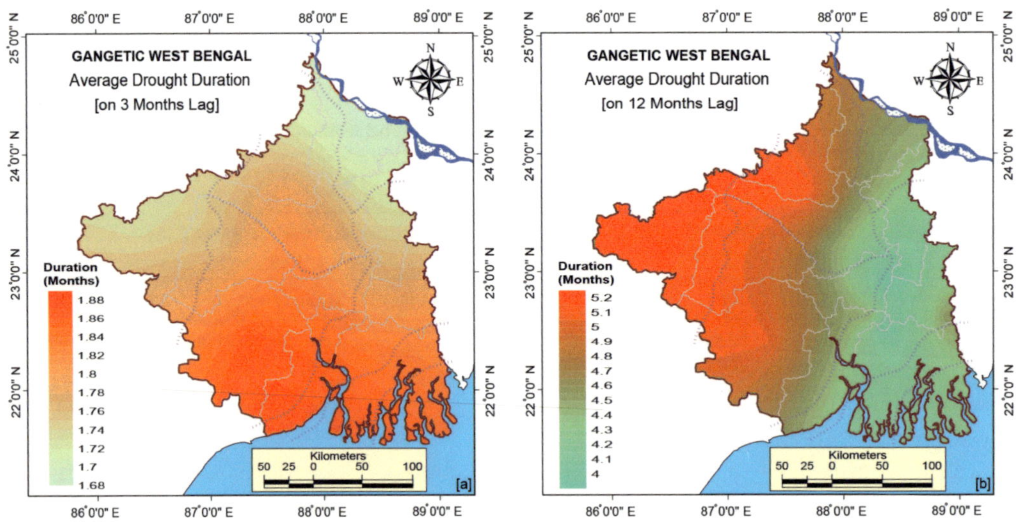

Figure 6. (a) Average duration of drought events for (a) 3-, (b) 12-month time scale.

4.3. Assessment of the frequency of drought occurrences

4.3.1. Regional average frequency

Percentage frequencies of drought occurrences of varying drought categories at different time steps have been outlined in **Table 4**. At all the time steps, more or less 16% years (i.e. once in every 6 year) have recorded drought of all categories.

4.3.2. Spatial character of the frequency of drought occurrences

At the 12-month time steps the moribund delta followed by parts of northern as well as southern Rarh has experienced higher occurrence severe droughts (**Figure 8a**). Extreme drought occurrences, on the other hand, are more pronounced in the western degraded plateau region (**Figure 8b**). This means that the western degraded plateau suffers from extreme drought condition frequently while parts of Rarh Bengal and northern extreme of the deltaic Bengal suffer from frequent severe drought conditions.

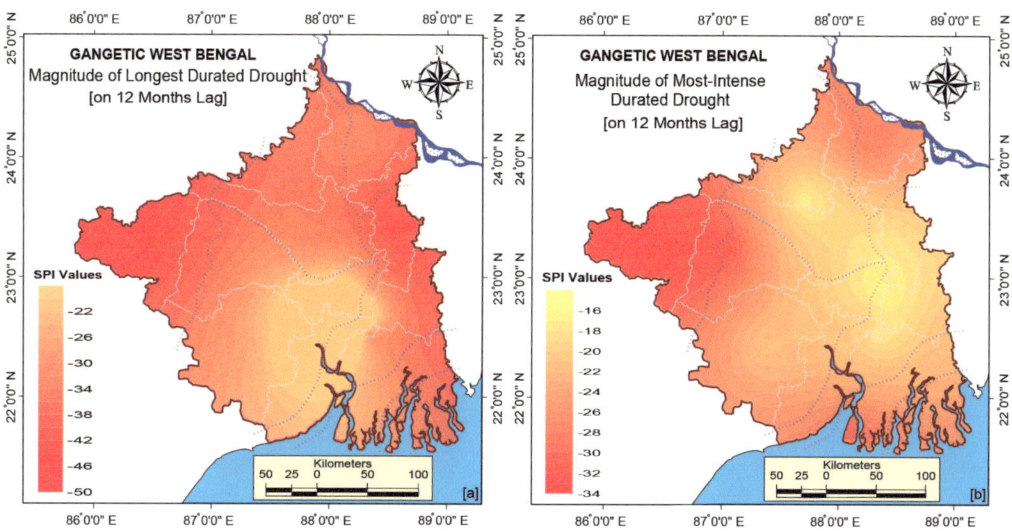

Figure 7. Magnitude for (a) longest and (b) most most-intense duration draughts on 12-month time step.

Draught severity class	SPI values	3-M lag	12-M lag	24-M lag	48-M lag
Moderately dry	(−1.0 to −1.49)	8.76	10.80	7.83	13.08
Severely dry	(−1.5 to −1.99)	4.09	4.20	5.66	3.91
Extremely dry	(−2 and less)	3.11	1.24	2.66	1.10
Total (%) of all categories drought		15.96	16.24	16.15	18.10

Table 4. Frequency of drought occurrences (%) in GWB at different time steps.

4.4. Phase-wise pattern of drought intensity, duration and frequency

To evaluate the Phase-wise change of drought variables during 1901–2002, the annual rainfall data series have been fitted with LOWESS or locally weighted regression curves [33, 34] to identify the patterns over time (**Figure 9**) and thereby to divide the entire time span into some clearly distinct phases (**Table 5**).

Regional average as well as peak drought intensity has increased significantly during first two consecutive phases. Meanwhile, there is no obvious change in average and maximum drought intensity between phase-II and Phase-III. The average drought duration has increased from 2.6 months in the phase-I to 5.5 month in the Phase-II and again has decreased to 4.4 months in the phase-III. Meanwhile, the most intense duration has increased from 8 month (Phase-I) to 11 month (Phase-II) again has equalized to 7 months in Phase-III. This may seem to be a good sign, but there is significant increase in mean intensity in the most intense duration drought from −1.46 in phase-I to as maximum as −2.06 in phase-III. This signifies that extreme drought events become short duration but its intensity is escalating as a signature of climate change. The regional extreme drought frequency during 1965–2002 compared to 1901–1933 has increased from 0 to about 2%.

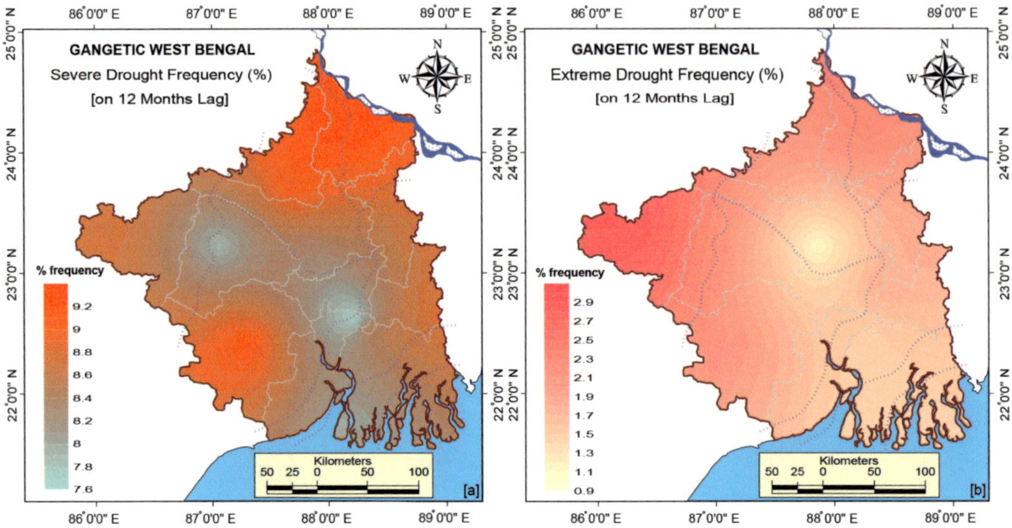

Figure 8. (a) Severe and (c) extreme drought occurrences at 12-month time steps.

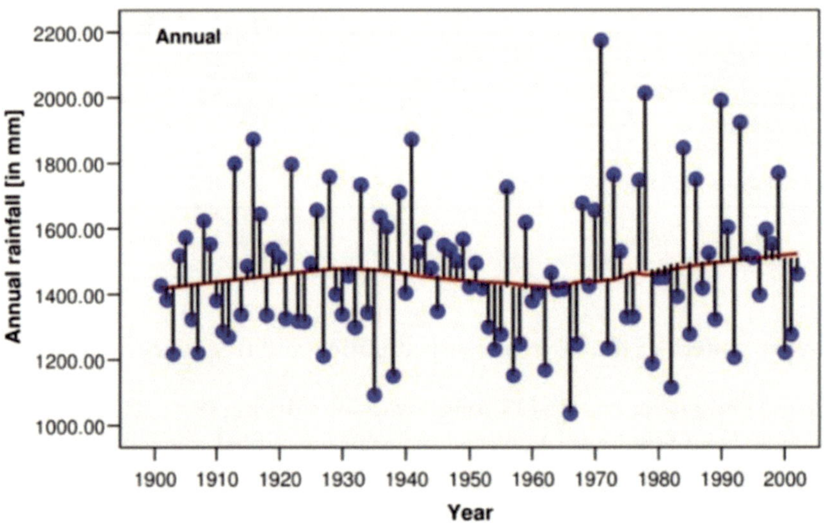

Figure 9. Locally weighted regression (LOWESS) and scatter plots of the average annual rainfall in GWB for the period 1901–2002.

4.5. Drought trend assessment

The MK test has been used to identify trend of drought intensity (**Table** 6). Here, in case of drought analysis, positive trend will indicate intensification of wet condition and negative change will indicate amplification of dry condition because trend of drought intensity is inversely related to the rainfall trend i.e. if rainfall declines drought event will amplify.

Drought indicators	Phase-I: 1901–1933	Phase-II: 1934–1964	Phase-III: 1965–2002	1901–2002
Average drought intensity (MID)	−0.71	−0.82	−0.77	−0.77
Maximum drought intensity (PID)	−1.98, 1903, July	−2.83, 1939, June	−2.82, 1967, June	−2.83, 1939, June
Average drought duration (DDA(M)) (SPI: ≤ −1.0 for consecutive months)	2.61	5.54	4.39	4.18 month
Maximum drought duration (DDL(M)) (SPI: ≤ −1.0 for consecutive months)	10 month (August, 1927 to May, 1928)	21 month (Sept, 1957 to May, 1959)	14 month (July, 1966 to August, 1967)	21 month (Sept, 1957 to May, 1959)
Most intense duration (DDI(M)) (SPI: ≤ −2.0 for consecutive months)	8 month (July, 1903 to Feb, 1904). Mean intensity: −1.46	11 month (Sept, 1938 to July, 1939). Mean intensity: −1.68	7 month (Sept, 1966 to March, 1967). Mean intensity: −2.06	7 month (Sept, 1966 to March, 1967): Mean intensity: −2.06
Moderate drought frequency (%)	10.65	10.65	12.73	10.80
Severe drought frequency (%)	2.69	6.72	4.30	4.20
Extreme drought frequency (%)	0	1.32	1.97	1.24

Table 5. Regional intensity, duration and frequency of drought events identified from SPI values at a 12-month scale for different periods in GWB.

At all the time steps, stations Berhampore, Purulia, Bankura, Sriniketan and Burdwan, irrespective of their level of significance, have experienced amplification of dry condition over the assessed period. For the 3-month step, significant positive trends (on 95% sig. level) have been detected at Berhampore, Sriniketan and Purulia of the Moribund delta, northern Rarh and western degraded plateau respectively (**Figure 10a**). For the station Sriniketan and Purulia the trend remains analogous for the 12- and 24-month time series also (**Figure 10b** and **Table 6**). However, on longer time span of 24-month scale, station Berhampore of the Moribund deltaic Bengal has experienced insignificant growth on 95% level of significance unlike the 3- and 12-month scale (**Table 6**).

4.6. Drought returns periods

In case of mod drought, the return periods are lesser in parts of deltaic Bengal and Rarh plain (**Figure 11a**). However, the circumstances roughly reversed in case of the extreme drought (**Figure 11b**) and the western degraded plateau and some parts of the plateau fringe fans and moribund delta part return periods count lesser. Thus, on entire regional scale the deltaic Bengal & Rarh plain are more sensitive to frequent attack of severe drought but extreme drought attack more recurrently in the western degraded plateau and some adjacent parts of the Rarh plain.

Weather stations	SPI-3			SPI-12			SPI-24		
	Z	Q	Tr	Z	Q	Tr	Z	Q	Tr
Berhampore	-1.30^{*}	−0.21	In^{*}	-2.05^{*}	−0.49	In^{*}	-1.87^{+}	−0.59	In^{+}
Krishnanagar	0.44	0.07	De	0.94	0.22	De	1.49	0.39	De
Chinsurah	1.17	0.17	De	1.86	0.48	De	2.52^{+}	0.75	De^{+}
Uluberia	1.75^{+}	0.24	De^{+}	2.77^{+}	0.71	De^{+}	3.23*	0.99	De*
Alipore	1.46^{+}	0.22	De^{+}	2.40^{+}	0.61	De^{+}	2.98*	0.91	De*
Basirhat	1.98*	0.32	De*	3.27*	0.81	De*	3.88*	0.15	De*
Purulia	−1.26	−0.18	In^{+}	-1.79^{+}	−0.44	In^{+}	-1.95^{+}	−0.60	In^{+}
Bankura	−0.53	−0.10	*In*	−0.80	−0.18	*In*	−0.63	−0.20	*In*
Sriniketan	−1.49*	−0.24	In^{*}	-2.53^{*}	−0.63	In^{*}	−2.77*	−0.85	*In**
Burdwan	−0.45	−0.06	*In*	−0.81	−0.18	*In*	−0.61	−0.16	*In*
Midnapore	1.33	0.20	De	2.22	0.55	De	2.81^{+}	0.80	De^{+}
Sagar Island	2.17*	0.34	De*	3.48*	0.89	De*	4.00*	0.25	De*
GWB Avg.	0.53	0.08	De	0.92	0.24	De	1.56	0.44	De

Note: ***Z*:** standardized test statistics of MK test; ***Q*:** Sen's slope estimate; ***Tr*:** trend of drought; ***In*:** increasing and ***De*:** decreasing trend of dry condition. *Significant trend at 0.05 level of significance.
$^{+}$Significant trend at 0.1 level of significance.

Table 6. Result of MK test, Sen's slope and trend of drought (1901–2002) over GWB.

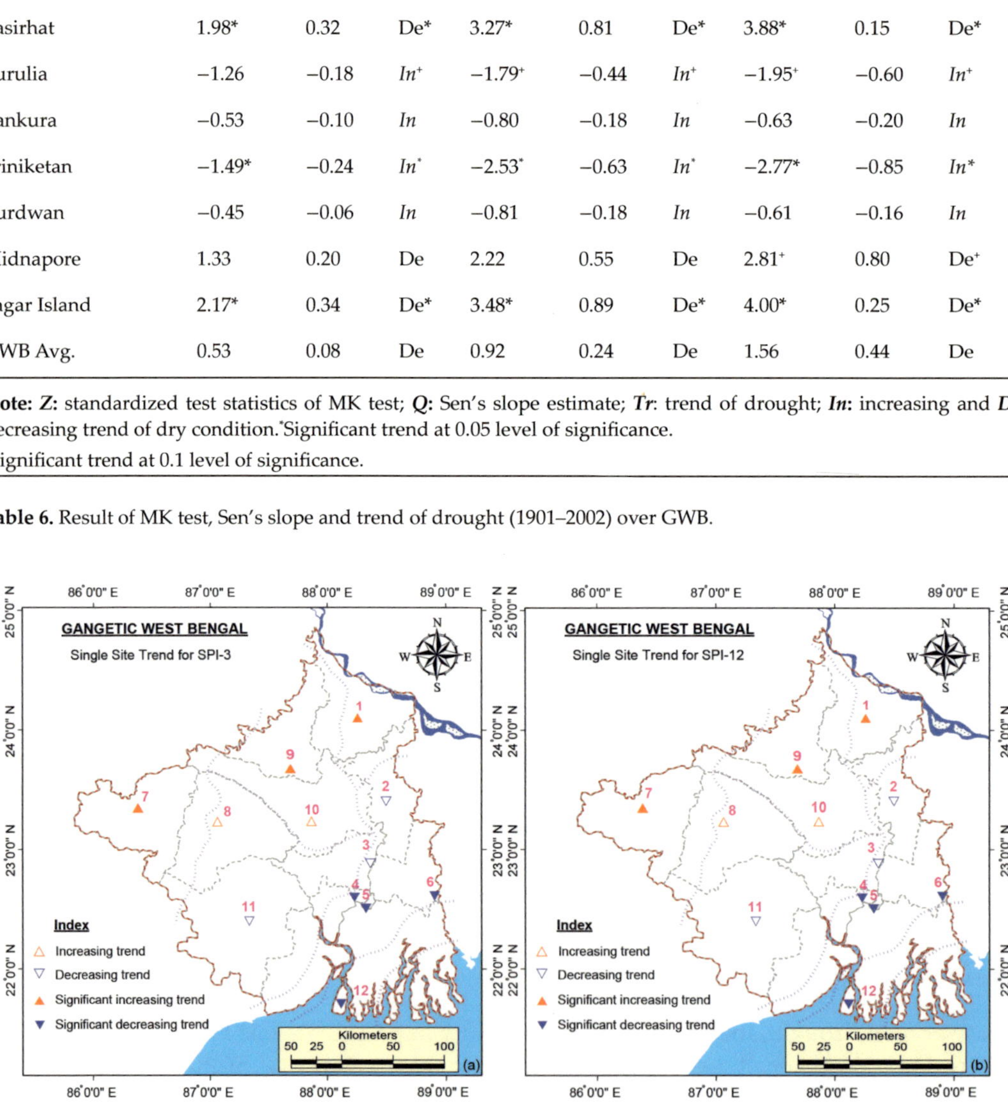

Figure 10. Single-site trend of drought intensity for SPI at different time scales.

http://dx.doi.org/10.5772/intechopen.80037

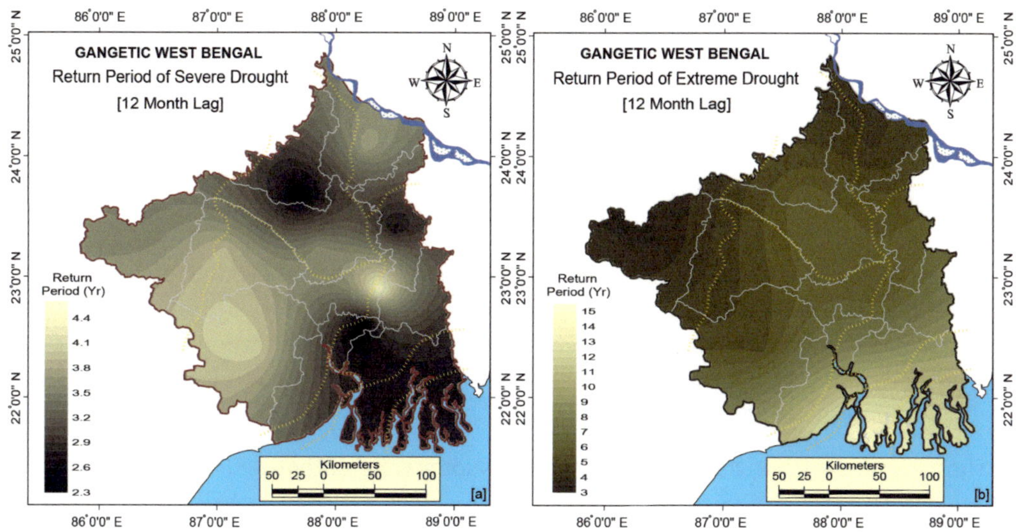

Figure 11. Return period of (a) severe drought (SPI < −1.5 to −1.99) and (b) extreme drought (SPI < −2.0) on 12-month time steps.

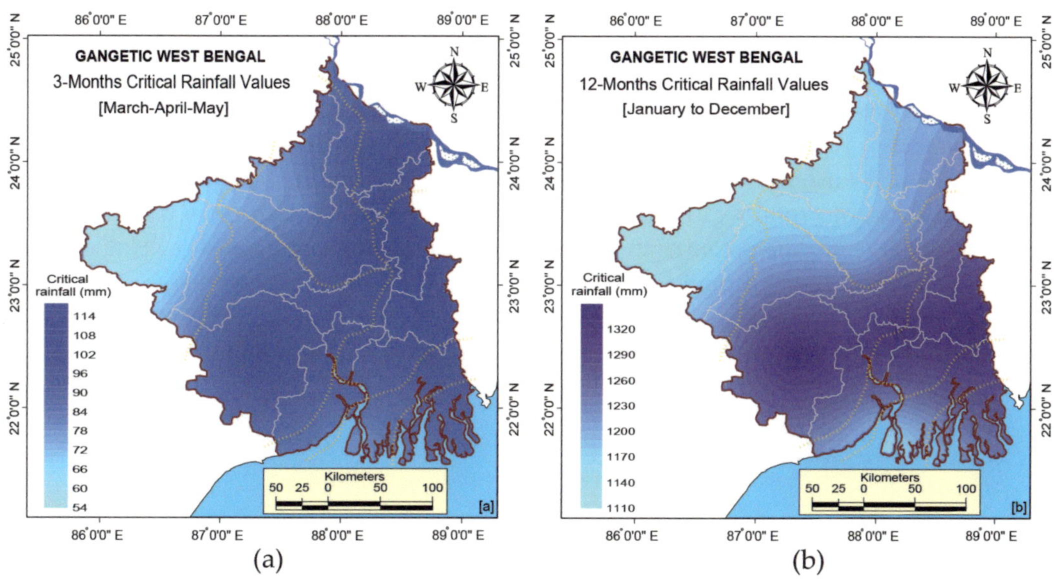

Figure 12. Critical rainfall values for (a) 3-months and (b) 12-month time lag.

4.7. Critical (threshold) rainfall analysis

Critical rainfall is the least amount of rainfall below which can initiate drought. The critical rainfall demands for no-drought occurrences roughly increases from West to East and

South-east (**Figure 12a** and **b**). Therefore, it is likely that the interior region immediately after the coastal and lower Ganga delta will be more exposed to droughts particularly the non-irrigated croplands due to relatively more demand of water.

5. Summary and conclusion

This chapter has provided results on the assessment meteorological drought condition for the Gangetic West Bengal of Eastern India over the last century in the context of climate change. The patterns of drought frequency, magnitude, trend etc. portrayed through statistical assessment, visually interpretive maps and geographic description so as to improve our understanding of drought jeopardy of this region.

The study confirmed that, the last century exhibits some consecutive deficit and surplus phases and after 1950s extremity of surplus and deficit as well as duration have been increased substantially. Stepping up of the maximum drought intensity; mean intensity of the most intense drought event; average drought duration; severe and extreme drought frequency from 1940s in this agricultural tract are some alarming events to think over. At the intra-regional scale, average drought intensity as well as duration was the greatest in the western degraded plateau and Rarh region and are also sensitive to extreme droughts. The impact of drought is expected to be rigorous at or adjacent areas of the western degraded plateau, particularly the northern Rarh and moribund delta where the drought intensities are tend to increase while the rainfall as well as recurrence interval of drought are tend to lessen. The western degraded plateau is widely known for its drought proneness [35]. But, this work provides evidences demonstrating the extension and intensification of dryness at and in the adjacent areas of this traditional drought prone region namely, towards the northern Rarh plain and moribund delta. All of these indicate potential threat to the rain-fed agriculture, food security and socio-economic vulnerability to drought of this region. Therefore, a more detailed study to explore the drought risk as well as trend and pattern of other hydro-climatic variables is essential. Work related to this issue is in progress and will be reported elsewhere.

Acknowledgements

The author is like to express his sincere thanks to Prof. Sutapa Mukhopadhyay of the Department of Geography, Visva-Bharati University, Santiniketan and Dr. Swades Pal of the Gour Banga University, Malda for their fair and priceless assistance in upgrading the present piece of writing. The author is also appreciative to the India Meteorological Department for providing the data.

Conflict of interest

The author states that there is no conflict of interest.

Author details

Krishna Gopal Ghosh

Address all correspondence to: k.g.ghosh@india.com

Department of Geography, Malda College, Malda, West Bengal, India

References

[1] Science Daily. Asia's Most Devastating Droughts Reconstructed. The Earth Institute at Columbia University; 2010. www.sciencedaily.com/releases/2010/04/100422153929.htm [Accessed: 13 May 2016]

[2] Widawasky DA, O'Toole JC. Prioritizing the rice biotechnology research agenda for Eastern India. New York, NY: Rockefeller Foundation Press; 1990. 86 p

[3] Miyan MA. Droughts in Asian least developed countries: Vulnerability and sustainability. Weather and Climate Extremes. 2015;**7**:8-23. DOI: 10.1016/j.wace.2014.06.003

[4] Kamble MV, Ghosh K, Rajeevan M, Samui RP. Drought monitoring over India through normalized difference vegetation index (NDVI). Mausam. 2010;**61**:537-546

[5] INCCA: Indian Network for Climate Change Assessment. Climate Change and India: A 4×4 Assessment a Sectoral and Regional Analysis for 2030s. Ministry of Environment & Forests Government of India; 2010. 164 p. Available from: www.moef.nic.in/downloads/public-information/fin-rpt-incca.pdf [Accessed: 13 May 2016]

[6] WBSAPCC: West Bengal State Action Plan on Climate Change. Government of West Bengal, India; 2010. http://moef.nic.in/downloads/public-information/West-Bengal-SAPCC.pdf [Accessed: 30 November 2013]

[7] Mishra S. Abar Kharar Kabale Paschim Banga (in Bengali). Saar Samachar. 2010;**48**(3):11-30

[8] RPAPCC: Regional Policy Action Platform on Climate Change. Climate Change Policy Paper I-IV; 2012. http://www.wwfindia.org/?7560/regional-policy-action-platform-on-climate-change-rpapcc-policy-paper [Accessed: 19 November 2015]

[9] Ghosh KG. Spatial and temporal analysis of rainfall trend over Gangetic West Bengal, Eastern India in the context of climate change. Landscape. 2016;**2**:73-96

[10] Ghosh KG. Long range climatic variability over Birbhum District, West Bengal and their Impact on Rainfed Aman Crop in the Context of Climate Change: Adoption and Mitigation. In: Chattopadhyay PK, Kushwaha DS, editors. Rural Health, Women empowerment and Agriculture: Issues and Challenges. 1st ed. India: New Delhi Publishers; 2016. Chap. 21. pp. 277-298

[11] Biswas KR. Rivers of Bengal: A Compilation. In: A Series of 5 Volume edited by West Bengal District Gazetteers, Higher education Department, Govt of West Bengal, Kolkata, India. 2001

[12] Mishra S. Weather and climate of West Bengal. Government of West Bengal, June, XLVIII. 2006;**6**:17-28. http://westbengal.gov.in/BanglarMukh/Download?FilePath=/alfresco/d/d/workspace/SpacesStore/1111a108-a276-4849-9719-38629b02144f/Chap-Wcw_17.PDF Accessed 23 Dec 2013

[13] AD-GoWB: Agriculture Department, Govt of West Bengal. State Agricultural Plan for West Bengal. 2009. http://www.rkvy.nic.in/static/SAP/WB/WB.PDF Accessed 3 November 2015

[14] Bandyopadhyay S, Kar NS, Das S, Sen J. River Systems and Water Resources of West Bengal: A Review. Geological Society of India Special Publication. 2014;**3**:63-84

[15] DoSPI-GoWB: Department of Statistics and Programme Implementation, Govt. of West Bengal. In: Economic Review: 2011-12, Bureau of Applied Economic Statistics. Kolkata; 2012. p. 305

[16] Nath SK, Roy D, Thingbaijam SKK. Disaster mitigation and management for West Benga. India-An appraisal. Current Science. 2008;**94**(4):858-864

[17] Dracup JA, Lee KS, Paulson EG Jr. On the definition of droughts. Water Resource Research. 1980;**16**(2):297-302. DOI: 10.1029/WR016i002p00297

[18] McKee TB, Doesken NJ, Kleist J. The relation of drought frequency and duration to time scales. Proceeding of the Eight Conference on Applied Climatology. 17-22 January, Anaheim, California. Amer Meteor Soc. Boston, Massachusetts; 1993. pp. 179-184

[19] McKee TB, Doesken NJ, Kleist J. Drought monitoring with multiple time scales. Proceeding of the Ninth Conference on Applied Climatology. 15-20 January, Dallas, Texas. Amer Meteor Soc. Boston, Massachusetts; 1995. pp. 233-236

[20] WMO: World Meteorological Organization. In: Svoboda M, Hayes M, Wood D, editors. Standardized Precipitation Index User Guide. WMO-No 1090, Geneva; 2012 http://www.wamis.org/agm/pubs/SPI/WMO_1090_EN.pdf Accessed 23 November 2015

[21] Sonmez FK, Komuscu AU, Erkan A, Turgu E. An analysis of spatial and temporal dimension of drought vulnerability in Turkey using the Standardized Precipitation Index. Natural Hazards. 2005;**35**(2):243-264. DOI: 10.1007/s11069-004-5704-7

[22] Morid S, Smakhtin V, Moghaddasi M. Comparison of seven meteorological indices for drought monitoring in Iran. International Journal of Climatology. 2006;**26**:971-985. DOI: 10.1002/joc.1264

[23] Damberg L, AghaKouchak A. Global trends and patterns of drought from space. Theoretical and Applied Climatology. 2014;**117**:441-448. DOI: 10.1007/s00704-013-1019-5

[24] Dupigny-Giroux LA. Towards characterizing and planning for drought in Vermont: Part I. A climatological perspective. Journal of the American Water Resources Association. 2001;**37**(3):505-525. DOI: 10.1111/j.1752-1688.2001.tb05489.x

[25] Spinoni J, Naumann G, Carrao H, Barbosa P, Vogt J. World drought frequency, duration, and severity for 1951-2010. International Journal of Climatology. 2014;**34**(8):2792-2804. DOI: 10.1002/joc.3875

[26] Thompson S. Hydrology for water management. Rotterdam, The Netherlands: AA Balkema Publication; 1999. p. 476

[27] Wang QF, Wu JJ, Lei TJ, He B, Wu ZT, Liu M, Mo XY, Geng GP, Li XH, Zhou HK, et al. Temporal-spatial characteristics of severe drought events and their impact on agriculture on a global scale. Quaternary International. 2014;**349**:10-21. DOI: 10.1016/j.quaint.2014.06.021

[28] Mann HB. Nonparametric tests against trend. Econometrica. 1945;**13**:245-259

[29] Kendall MG. Rank Correlation Methods. London, UK: Charles Griffin; 1975

[30] Sen PK. Estimates of the regression coefficient based on Kendall's tau. Journal of the American Statistical Association. 1968;**63**(324):1379-1389. DOI: 10.1080/01621459.1968.10480934

[31] Bonaccorso B, Bordi I, Cancielliere A, Rossi G, Sutera A. Spatial Variability of Drought: An Analysis of the SPI in Sicily. Water Resources Management. 2003;**17**(4):273-296. DOI: 10.1023/A:1024716530289

[32] Haan CT. Statistical Methods in Hydrology (ed). Iowa, USA: The Iowa State University Press; 1977. p. 378

[33] Cleveland WS. Robust Locally Weighted Regression and Smoothing Scatterplots. Journal of the American Statistical Association. 1979;**74**(368):829-836. DOI: 10.1080/01621459.1979.10481038

[34] Cleveland WS. Graphs in Scientific Publications. Amer. Statist. 1974;**38**(4):261-269. DOI: 10.1080/00031305.1984.10483223

[35] WBPCB: West Bengal Pollution Control Board. A State of Environment Report: Water Resource and its Quality in West Bengal, Kolkata; 2009. 352 p

Chapter 6

Rainfall Erosivity: Gap-Filling Method Differences in the Brazilian Pantanal and Cerrado Biomes

Diego A. Zanoni, Susana P. Moreira,
Ana Paula S. Teles, Guilherme H. Cavazzana,
Denilson O. Guilherme and
Fernando JC. Magalhães Filho

Additional information is available at the end of the chapter

http://dx.doi.org/10.5772/intechopen.77064

Abstract

To improve the use of soil and its conservation, precipitation data are necessary. With the Universal Soil Loss Equation (USLE), the study of historical precipitation series is a main factor, but in these series, there are gaps that need to be filled. This study had, as a basis, the methods of weighted likelihood, multiple regression, and weighted likelihood based on multiple regression to fill the gaps of the rainfall data for the rainfall gauges in the Brazilian biomes (Cerrado and Pantanal, municipalities of Campo Grande, Bandeirantes, Sidrolândia, Miranda, Fazenda Ponte, and Ribas do Rio Pardo). With this, it became possible to calculate the rainfall erosivity (R factor in the USLE). Therefore, the consistency of the filled rainfall data was analyzed by the double mass method. The value of the rainfall erosivity calculated varies from 2304.80 to 13562.10 MJ mm ha^{-1} h^{-1} year. With this data, it was possible to identify particular results that differed from the rainfall erosivity. Comparing all the gap-filling methods, numbers varying from 0–12% at the same rainfall gauge were obtained.

Keywords: water and soil conservation, USLE, water resources

1. Introduction

Climate changes are each day more and more notable throughout the world, and based on this fact, scientific studies are being developed, having as one of the main subjects studied being rainfall and its historical series [1].

In studying rainfall and its historical series, gaps in its data (these gaps can occur due to equipment failure or data observer's mistakes, which are the most common reasons) were found. These gaps can occur in hourly, monthly, or annually collected data. In some situations, these gaps make it impossible to use the data in some studies [2].

With the need of filling these gaps, some methods were developed and are often used in studies. These methods include the artificial neural network (ANN) method, as it can be seen in [3–5], weighted likelihood method [6, 7], multiple regression method [8, 9], and weighted likelihood based on multiple regression method [1, 2].

Consequently, with the development of these methods, it was necessary to create a method to analyze the data consistency when its gaps were filled. According to this, [10] developed a method called double mass.

A continuous historical rainfall series, with filled gaps and analyzed consistency, can be applied in many studies such as urban drainage, soil conservation, and water conservation. In the soil conservation field, many studies have been developed about soil loss due to water erosion [11–13], which is described by the equation proposed by [14], which considers variables like the soil erodibility, topographic factor, soil use and management, conservation practices, and rainfall erosivity.

The rainfall erosivity (Rc) is calculated based on historical rainfall series, and to obtain these continuous historical series, certain methods are used, where the resultant data can be different depending on the used method. Based on this, this study was developed aiming to analyze the differences obtained in the rainfall erosivity results calculated with filled rainfall data using the methods—weighted likelihood, multiple regression, and weighted likelihood based on multiple regression—and to obtain a better correlation coefficient between different hydrological data sources (radar, satellite, and local).

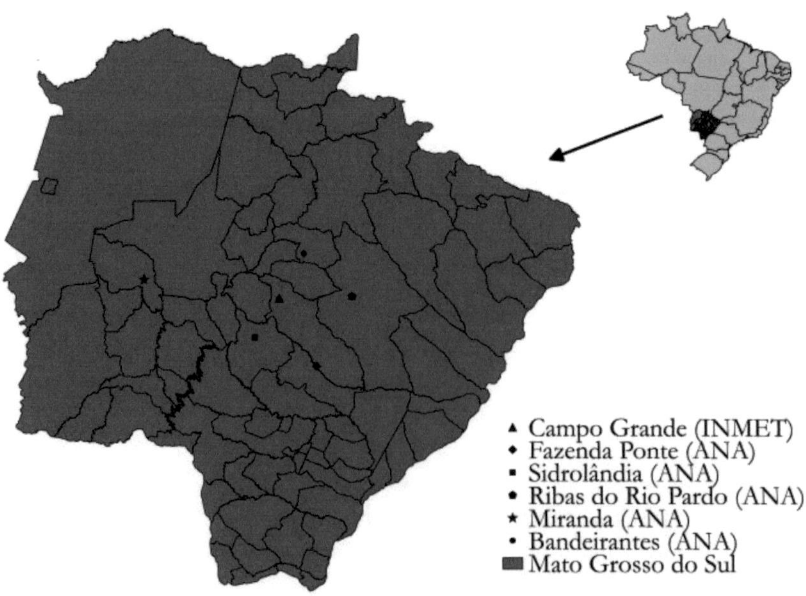

Figure 1. Rainfall gauges.

2. Area of study

In order to fill the gaps in historical series, an auxiliary rainfall gauge is used. For each gap in the series, it is advised to use at least three other values from an auxiliary rainfall gauge [2]. Therefore, six rainfall gauges (**Figure 1**) were used: Campo Grande, Bandeirantes, Sidrolândia, Miranda, Fazenda Ponte, and Ribas do Rio Pardo. These stations are located in the Paraná River Basin, in the central area of the state of Mato Grosso do Sul, Brazil. With the stations defined, values of historical series from the Agência Nacional de Águas (ANA) and Instituto Nacional de Meteorologia (INMET) data base were collected. The range with the fewest gaps was from January 1, 2001, to December 31, 2014, and for filling the gaps, three methods were used.

3. Methods used to fill gaps

3.1. Weighted likelihood method

As it can be seen in [2], the month without data is filled with Eq. (1):

$$P_x = 1/n \sum_{i=1}^{n} N_x/N_i \; P_i \tag{1}$$

where:

P_x = data to be filled (mm),

n = number of auxiliary stations,

N_x = annual average rainfall at the station without data (mm),

N_i = annual average rainfall at the auxiliary station (mm),

P_i = rainfall at the auxiliary station in the month to be filled (mm).

3.2. Multiple regression method

As pointed out in [15], the multiple regression method is based on applying multiple regression, establishing a relation among the auxiliary stations and the station with the data gap, and it uses Eq. (2):

$$y_c = x_{1i} + a_1 x_{2i} + \cdots + a_{n-1} x_{ni} + a_n \tag{2}$$

where:

y_c = data to be filled (mm).

n = number of auxiliary stations.

a_n = coefficient to be estimated by multiple regression.

x_{1i}, x_{2i} = data in auxiliary stations (mm).

3.3. Weighted likelihood based on multiple regression (mixed)

As it can be seen in [16], weighted likelihood based on multiple regression is based on mixing both of the previous methods. To make it easier to understand the name of this method, it will be denoted as mixed. First a multiple regression between the gap station and each auxiliary station is calculated separately, and then the weight of each station related to the gap station is calculated, using Eq. (3):

$$W_{yxj}=r_{yxj}/\left(r_{yx1}+r_{yx2}+\cdots+r_{yxn}\right) \tag{3}$$

where:

W_{yxj} = weight factor between the station (y, gap station, and xj auxiliary station),

r_{yxj} = correlation coefficient between the gap station and auxiliary station (linear regression coefficient),

n = number of auxiliary stations.

For each auxiliary station, a value of W_{yxj} was obtained. With all the weight factors calculated, their sum must be equal to 1. After this is calculated, the data for the gap is calculated with Eq. (4):

$$y_c = x_1 W_{x1}+x_2 W_{x2}+\cdots+x_n W_{xn} \tag{4}$$

where:

y_c = station data to be filled (mm),

x_n = rainfall data at the auxiliary station in the month of the data to be filled (mm),

W_{xn} = weight factor between gap station and auxiliary station.

3.4. Double mass method

With the data gaps filled by all three methods, it was necessary to analyze the data consistency. To achieve this, the double mass method was used as it can be seen in [17–19], described by [10], which consists of comparing two rainfall gauges using the amount of rainfall during the period of the study, developing a chart using a reliable station on the x-axis and the station to be compared on the y-axis. The chart developed tends to be a straight line; the inclination of the straight line represents the correlation between the reliable station and the station to be analyzed.

In some cases, the dots of the chart may not tend to be a straight line, or it may appear as ranges in the chart with different inclinations from the straight line found in the linear regression. These differences may be because of the failure in obtaining the data, changing the observer at the station, changing the environment nearby the station, or changing the location of the monitoring equipment [10].

To analyze the consistency, the station in Sidrolândia was determined as reliable and, therefore, used as a reference, since during the period of study, there was only one value gap.

4. Rainfall erosivity

To obtain the value of rainfall erosivity, Eq. (5) was used described by [20]

$$R_c = P^{\wedge 2}/P \quad (5)$$

where:

Rc = rainfall coefficient (mm),

p^2 = monthly rainfall (mm),

P = annual average rainfall (mm) according to [21] who has defined these parameters in Eq. (5) to be applied in Campo Grande, as it can be seen in [11]. These parameters are shown in Eq. (6):

$$R=139{,}44\times(p^{\wedge 2}/P)^{\wedge 0{,}6784} \quad (6)$$

where:

R = rainfall erosivity factor (MJ mm ha^{-1} h^{-1} year),

p = monthly rainfall,

P = annual average rainfall.

With this equation the rainfall erosivity for each method and for each rainfall gauge was calculated. After that, the rainfall erosivity for each station was classified for each method used. The following chart was used to classify the rainfall erosivity level (**Table 1**).

Erosivity (MJ mm ha^{-1} h^{-1})	Erosivity classification
R < 2.452	Low erosivity
2.452 < R < 4.905	Moderate erosivity
4.905 < R < 7.357	Moderate to high erosivity
7.357 < R < 9.810	High erosivity
R > 9.810	Very high erosivity

Source: [22], modified to I.S. metric of unity according to [23].

Table 1. Rainfall erosivity classification.

5. Gap filling

All the rainfall gauges had gaps in their historical rainfall series; with the total data having 156 values, the stations had the following gaps: Campo Grande 17.3%, Bandeirantes 3.2%, Sidrolândia 0.6%, Miranda 2.6%, Fazenda Ponte 1.9%, and Ribas do Rio Pardo 3.2%.

For the gap filling using the weighted likelihood method, the values were analyzed with the double mass method, and the dots close to the linear regression were satisfactory with R^2 values above 0.9936.

With the multiple regression method, the dispersion of the dots tended to be a straight-line regression with R^2 values above 0.9938.

And with the weighted likelihood based on multiple regression the dispersion of the dots tended to be a straight-line regression with R^2 values above 0.9938.

Even if the dispersion of the dots in all methods used had a satisfactory R^2 value, the double mass method shows consistency in all stations but for the Fazenda Ponte station, because the dispersion of the dots at this station for all methods showed a different inclination along the straight-line regression.

These differences are explained by [10]; these differences can happen because of many factors like changing the monitoring equipment operator, changing the environment nearby the station, or changing the location of the monitoring equipment.

6. Calculating the rainfall erosivity

With the results filled and the continuous historical rainfall series, the rainfall erosivity was calculated for each station and for each method. The values obtained can be seen in **Table 2**, for Campo Grande.

The period with higher rainfall erosivity at the Campo Grande station was in 2013 according to the weighted likelihood method and in 2011 for the other two methods. The year with the lowest rainfall erosivity was in 2002 and was equal for all methods. The consistency analysis can be seen in **Figure 2**.

It can be noticed that for all methods, the dispersion of the dots tends to be a straight-line regression; this implies that the inclination of all methods tends to be a straight line. The rainfall erosivity in Bandeirantes can be seen in **Table 3**.

For the Bandeirantes station, the years of maximum and minimum rainfall erosivity were the same for all methods, in 2011 and 2002, respectively. Furthermore, the amount of the annual rainfall erosivity differs 5% (between weighted likelihood and weighted likelihood based on multiple regression). The consistency analysis can be seen in **Figure 3**.

For the double mass comparison for all three methods, the same dispersion pattern of the dots can be noticed. They tend to be a straight-line regression, showing a consistency in the filled gaps. The rainfall erosivity in Sidrolândia can be seen in **Table 4**.

The year 2002 had the lowest rainfall erosivity for all methods, and 2003 had the highest rainfall erosivity, also for all methods. The rainfall erosivity in Miranda can be seen in **Table 5**.

For the Miranda station, the period with the highest rainfall erosivity was in 2001 according to the weighted likelihood and multiple regression methods and in 2003 according to the weighted likelihood based on multiple regression method. The period with the lowest rainfall erosivity was in 2002 for all methods. The percentage difference between the weighted likelihood and weighted likelihood based on multiple regression methods was 10%. The double mass analysis can be seen in **Figure 4**.

Year	Weighted likelihood	Multiple	Mixed
2002	7390.3	7835.8	7498.0
2003	8699.5	9254.8	9953.1
2004	8115.8	8115.8	8115.8
2005	8988.5	9228.9	9045.0
2006	8054.3	8085.0	8059.2
2007	9680.6	9757.8	9696.3
2008	8394.8	8394.8	8394.8
2009	9849.2	9849.2	9849.2
2010	8437.2	8539.1	8479.4
2011	9795.6	10324.2	10134.4
2012	9507.7	9507.7	9507.7
2013	9857.1	9857.1	9857.1
2014	9480.0	9848.4	9671.3
Total	116,250	118,598	118,261
Dif. %	−2%	0%	2%
Maximum	9857.10	10324.20	10134.40
Minimum	7390.30	7835.80	7498.00
Average	8942.40	9123.00	9097.00

Table 2. Rainfall erosivity in Campo Grande.

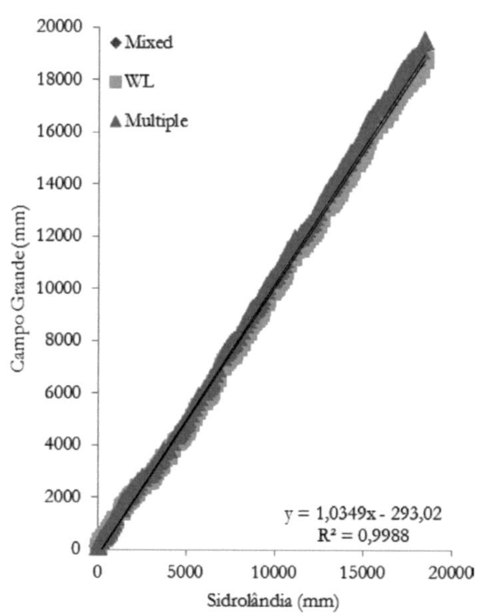

Figure 2. Campo Grande station analysis of double mass.

Year	Weighted likelihood	Multiple	Mixed
2002	7207.4	7207.4	6822.3
2003	9844.8	9844.8	10997.9
2004	8140.5	8140.5	8103.9
2005	9988.7	9988.7	10952.6
2006	9802.7	9802.7	11484.1
2007	8953.5	8953.5	8.317.3
2008	8655.5	8321.0	8894.2
2009	10440.0	10440.4	10807.4
2010	10214.0	10214.5	11654.5
2011	11396.0	11396.3	12128.6
2012	10469.0	10469.0	11168.3
2013	9492.0	9492.8	8812.0
2014	9268.0	9822.6	9864.4
Total	123,874	124,094	130007.6
Dif. %	0%	−5%	5%
Maximum	11396.3	11396.3	12128.6
Minimum	7207.40	7207.40	6822.30
Average	9528.80	9545.70	10000.60

Table 3. Rainfall erosivity in Bandeirantes.

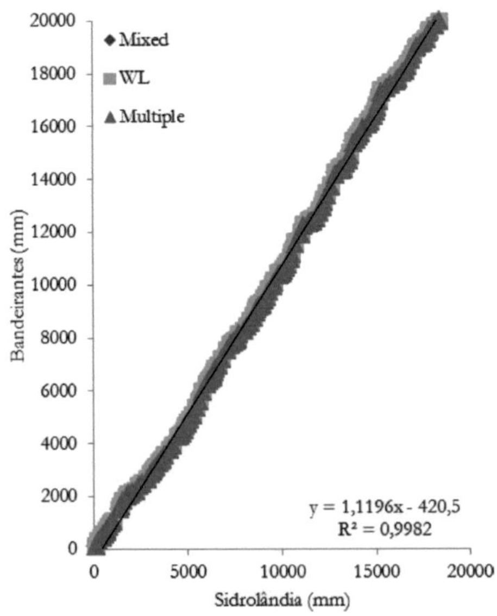

Figure 3. Bandeirantes station analysis of double mass.

Year	Weighted likelihood	Multiple	Mixed
2002	6959.2	6959.2	6430.5
2003	10078.8	10078.8	11049.6
2004	8525.7	8525.7	9011.1
2005	8479.7	8479.7	8133.8
2006	8043.5	8043.5	7797.5
2007	8676.2	8676.2	8401.6
2008	7330.7	7330.7	6718.1
2009	8521.0	8521.0	7545.5
2010	9491.9	9491.9	10599.2
2011	8927.3	8927.3	8170.2
2012	8607.5	8607.5	7826.6
2013	9686.7	9686.7	9271.5
2014	9513.2	9688.3	10015.3
Total	112,841	113,016	110,970
Dif. %	0%	2%	−2%
Maximum	10078.80	10078.80	11049.60
Minimum	6959.20	6959.20	6430.50
Average	8680.10	8693.60	8536.20

Table 4. Rainfall erosivity in Sidrolândia.

The data consistency of the Miranda station can be analyzed through its dispersion of dots compared to a straight-line regression, showing the same inclination. The rainfall erosivity at the Fazenda Ponte station can be seen in **Table 6**.

The period with the highest rainfall erosivity at the Fazenda Ponte station was in 2007 (weighted likelihood and multiple regression) and in 2003 (weighted likelihood based on multiple regression), and the period with lowest rainfall erosion was in 2008 (weighted likelihood and multiple regression) and in 2009 (weighted likelihood based on multiple regression). For this method the largest percentage difference was 12%. The double mass analysis can be seen in **Figure 5**.

Analyzing the dispersion of the data dots from this station, it is noticed that in some parts, the dispersion of the dots has different inclinations along the chart, even if the statistic approach is satisfactory ($R^2 = 0.9936$). These differences along the straight-line regression show us inconsistencies described by [10]. These different inclinations can be explained by facts such as changes of monitoring station operators, environmental changes nearby the location of the station, and changing of the location of the monitoring equipment. The rainfall erosivity in Ribas do Rio Pardo can be seen in **Table 7**.

Year	Weighted likelihood	Multiple	Mixed
2002	6221.7	6221.7	4736.0
2003	8780.3	8780.3	9272.5
2004	8533.0	8533.0	8798.5
2005	9185.4	9185.4	8589.9
2006	6520.9	6566.2	5745.0
2007	7139.4	7139.4	5382.3
2008	8047.4	8047.4	7246.6
2009	9359.5	9351.6	8434.3
2010	7978.7	7978.7	7239.7
2011	9506.9	9506.9	8547.2
2012	8201.3	8201.3	6987.4
2013	8160.9	8160.9	6650.1
2014	8142.1	8778.2	8415.1
Total	105,777	106,450	96,044
Dif. %	−1%	10%	−10%
Maximum	9506.90	9506.90	9272.50
Minimum	6221.70	6221.70	4736.00
Average	8136.70	8188.50	7388.10

Table 5. Rainfall erosivity in Miranda.

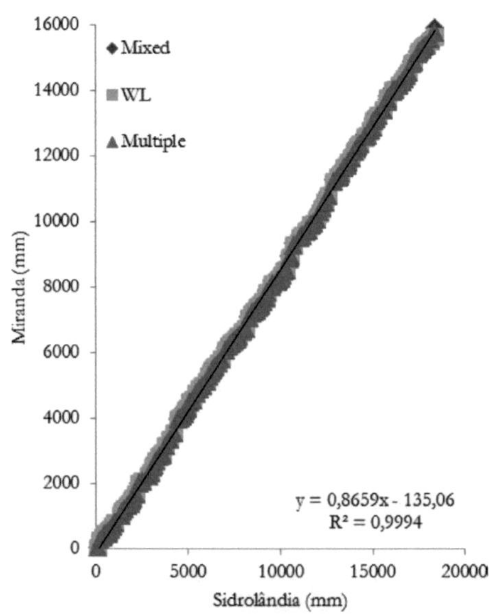

Figure 4. Miranda station analysis of double mass.

Year	Weighted likelihood	Multiple	Mixed
2002	8228.4	8228.4	8784.0
2003	8961.6	8961.6	10735.4
2004	5677.4	5677.4	3777.0
2005	8042.9	8042.9	6903.9
2006	7737.0	7886.2	8070.0
2007	9977.0	9977.0	8967.4
2008	5207.0	5341.9	3296.7
2009	5458.3	5458.3	2304.8
2010	9229.6	9229.6	9228.5
2011	9580.5	9580.5	8822.1
2012	8481.8	8481.8	7609.0
2013	7937.3	7937.3	6229.6
2014	7290.4	7290.4	5918.4
Total	101,809	102,093	90,647
Dif. %	0%	11%	−12%
Máximo	9977.0	9977.0	10735.40
Mínimo	5207.00	5341.90	2304.80
Média	7831.50	7853.30	6972.90

Table 6. Rainfall erosivity in Fazenda Ponte.

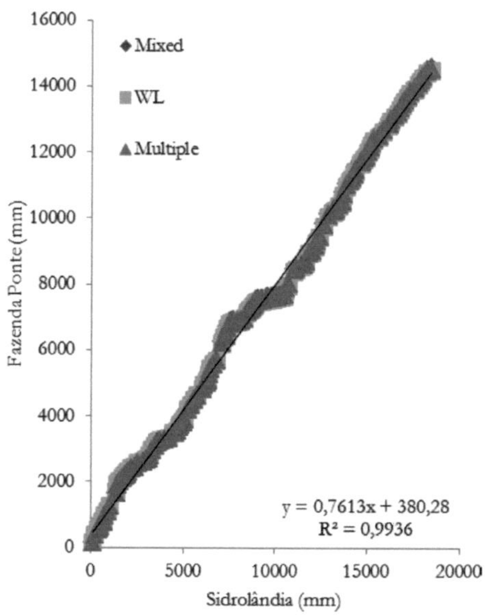

Figure 5. Fazenda Ponte station analysis of double mass.

Year	Weighted likelihood	Multiple	Mixed
2002	8564.3	8564.3	9544.8
2003	8549.5	8549.5	9251.4
2004	9306.9	9306.9	10365.7
2005	8835.9	8835.9	8506.0
2006	9246.2	9246.2	10475.0
2007	9750.6	9750.6	9898.2
2008	7252.5	6992.6	6352.9
2009	8998.5	8998.5	8213.7
2010	9541.0	9541.0	10217.7
2011	12165.6	12165.6	13562.1
2012	7957.2	7957.2	6592.5
2013	8270.9	8270.9	6724.7
2014	8506.9	8779.3	8773.6
Total	116,946	116,958	118,478
Dif. %	0%	−1%	1%
Máximo	12165.60	12165.60	13562.10
Mínimo	7252.50	6992.60	6352.90
Média	8995.90	8996.80	9113.70

Table 7. Rainfall erosivity in Ribas do Rio Pardo.

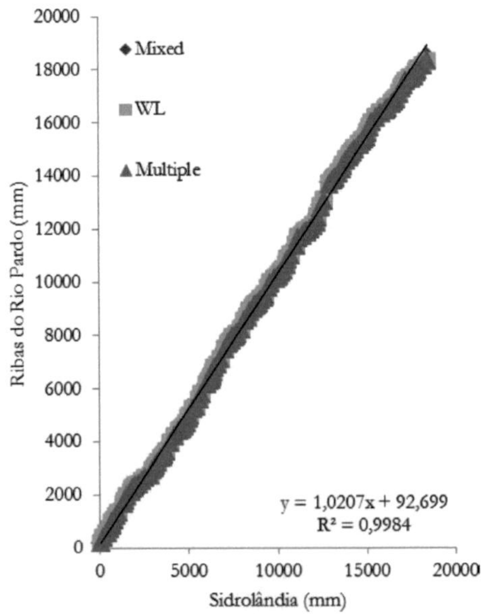

Figure 6. Ribas do Rio Pardo station analysis of double mass.

The period with the highest rainfall erosivity at the Ribas do Rio Pardo station was in 2013 and with the lowest rainfall erosivity was in 2008. This rainfall gauge had the lowest percentage difference among all the stations (1%). The double mass analysis can be seen in **Figure 6**.

The data of the Ribas do Rio Pardo station is noticed to be consistent; the dispersion of the dots tends to be a straight-line regression, with no change in inclination along the line.

The values of erosivity that were found in the study are in agreement with [24].

However, in the author's study, erosivity is classified only as high, and the gap filling can range from moderate to very high. This can be explained by the fact that the author [24] did not mention gap filling and used fewer stations in the study area (Cerrado and Pantanal biome).

7. Rainfall erosivity classification

After filling the gaps and analyzing the consistency of all rainfall gauges, the rainfall erosivity was classified according to **Table 1**. The rainfall erosivity in all stations was filled by the weighted likelihood method, which can be seen in **Table 8**.

The classification of rainfall erosivity using the multiple regression method to fill gaps can be seen in **Table 9**. The weighted likelihood based on multiple regression method and its classification results can be seen in **Table 10**.

After the classification was finished for each method, it can be noticed that the Bandeirantes and Fazenda Ponte stations had different rainfall erosivity classifications between the methods. Bandeirantes had a rainfall erosivity classification of very high erosivity (weighted likelihood mean based on multiple regression) and of high erosivity (weighted likelihood and multiple regression). The Fazenda Ponte station had a rainfall erosivity classification of moderate to high erosivity (weighted likelihood based on multiple regression) and high erosivity (weighted likelihood and multiple regression).

The results reported here may help to identify a more adequate methodology to fill the gaps in the region. It will subsidize appropriate plans and projects to infer the best land-use strategies to improve water and soil conservation and quality and promote agriculture sustainability.

Station	Average	Classification
Campo Grande	8942.4	High erosivity
Bandeirantes	9528.8	High erosivity
Sidrolândia	8680.1	High erosivity
Miranda	8136.7	High erosivity
Fazenda Ponte	7831.5	High erosivity
R. do Rio Pardo	8995.9	High erosivity

Table 8. Rainfall erosivity classification, using weighted likelihood method.

Station	Average	Classification
Campo Grande	9123.0	High erosivity
Bandeirantes	9545.7	High erosivity
Sidrolândia	8693.6	High erosivity
Miranda	8188.5	High erosivity
Fazenda Ponte	7853.3	High erosivity
R. do Rio Pardo	8996.8	High erosivity

Table 9. Rainfall erosivity classification, using multiple regression method.

Station	Average	Classification
Campo Grande	9097.0	High erosivity
Bandeirantes	10000.6	Very high erosivity
Sidrolândia	8536.2	High erosivity
Miranda	7388.1	High erosivity
Fazenda Ponte	6972.9	Moderate to high
R. do Rio Pardo	9113.7	High erosivity

Table 10. Rainfall erosivity classification, using weighted likelihood based on multiple regression method.

Mainly, this is important because the Cerrado is one of the Brazilian biomes that has been subjected to the highest agronomic pressure, according to [25], with high interaction with the Pantanal biome.

8. Rainfall data sources combined: satellite, radar, and local

When it is necessary to study rainfall, a couple of variables need to be determined, for example, the size of the area, duration of the historical rainfall series, and data source. Nowadays there are some kinds of sources available: satellite, radar, and local. How to choose among them? A very important detail when choosing the data source is to identify the size of the study area. When the study area is a state, a country, or a continent, the satellite data achieves a better accuracy. On the other hand, if the study area is a state or an area with a couple of cities, radar data is more advised. Finally, if the study area is a city, a small watersheds, or a couple of cities, the local data (rainfall gauges) provides a better result.

Satellite data is possible to be obtained using Precipitation Estimation from Remotely Sensed Information using Artificial Neural Networks-Climate Data Record (PERSIANN- CDR) [26], Tropical Rainfall Measuring Mission (TRMM) [27], and Gravity Recovery and Climate Experiment (GRACE) [28]. An important characteristic is the rainfall amount; some studies show that the same satellite data can provide different accuracy depending on the rainfall amount. Some characteristics are important to observe; in developing countries the number of local rainfall gauges is low, and some studies need to use two data sources combined.

Satellite data are widely used to calculate hydrological parameters for areas that are sparsely equipped with rain gauges; thus it is possible to obtain data for a large area. On the other hand, its accuracy for high rainfall quantities is low, for example, [26] describes a correlation coefficient of 0.62 between PERSIANN-CDR and data from local gauges, in a heavy rainfall event.

Consequently, different studies for each data set are required to obtain a better combination according to each study. This paper provides an approach to obtain more accurate data through different gap-filling methods. In this way, more studies are needed to sensor gap filling, providing future studies with bettered methods and combinations.

9. Conclusions and recommendations

1. The value of rainfall erosivity calculated varies from 2304.80 to 13562.10 MJ mm ha^{-1} h^{-1} year. It was possible to identify variations in rainfall erosivity classification, comparing all the gap-filling methods; numbers varying from 0–12% at the same rainfall gauge were obtained.
2. In the double mass analysis, even if statistic approaches are satisfactory and tend to be a straight-line regression, the inclination along the straight-line regression should be considered.
3. The consistency analysis can explain the different results obtained. The Fazenda Ponte station was an example where a break in the slope was found and the results obtained diverged 12%.
4. The weighted likelihood mean and multiple regression methods had similar performances in filling gaps; the rainfall erosivity values had a 2% maximum difference.
5. The weighted likelihood based on multiple regression was not often found in scientific articles even if it is adopted in books.
6. For future studies the use of the weighted likelihood based on multiple regression gap-filling method combined with a satellite data source is recommended.

Author details

Diego A. Zanoni, Susana P. Moreira, Ana Paula S. Teles, Guilherme H. Cavazzana, Denilson O. Guilherme and Fernando JC. Magalhães Filho*

*Address all correspondence to: fernando@ucdb.br

Dom Bosco Catholic University, Campo Grande, Brazil

References

[1] Ishihara JH, Fernandes LL, Duarte AAAM, Duarte ARCI, Ponte MX, Loureiro GE. Quantitative and spatial assessment of precipitation in the Brazilian Amazon (Legal Amazon) — (1978 to 2007). Revista Brasileira de Recursos Hídricos. 2014;**19**:29-39. DOI: 10.21168/rbrh.v19n1.p29-39

[2] Oliveira LFC, Fioreze AP, Medeiros AMM, Silva MAS. Comparison of gap filling methodologies of annual historical series of rainfall. Revista Brasileira de Engenharia Agrícola e Ambiental. 2010;**14**:1186-1192

[3] Detzel DHM, Oening AP, Souza ARR, Cerminaro SLC. Completing hourly river stage data via ARIMA models. Revista Brasileira de Recursos Hídricos. 2013;**18**:281-292

[4] Oliveira GG, Pedrollo OC, Castro NMR, Bravo JM. Hydrological simulations with different proportions of controlled area in the watershed. Revista Brasileira de Recursos Hídricos. 2013;**18**:193-204. DOI: 10.21168/rbrh.v18n3.p193-204

[5] Paiva LFG, Montenegro SMGL, Valença MJS. Runoff forecasting model based on the neutral networks technique for the São Francisco River basin. Revista Brasileira de Recursos Hídricos. 2010;**15**:05-15. DOI: 10.21168/rbrh.v15n1.p5-15

[6] Castro FS, Pezzopane JEM, Cecílio RA, Pezzopane JRM, Xavier AC. Evaluation of the performance of the different methods of interpolators for parameters of the climatologic water balance. Revista Brasileira de Engenharia Agrícola e Ambiental. 2010;**14**:871-880. DOI: 10.1590/S1415-43662010000800012

[7] Silva RM, Silva LP, Montenegro SMGL, Santos CAG. Analysis of the space-time variability and identification of rainfall pattern within the Tapecurá River basin, Pernambuco state. Sociedade and Natureza. 2010;**22**:357-372. DOI: 10.1590/S1982-45132010000200010

[8] Costa HC, Marcuzzo F, Ferreira OM, Andrade LR. Seasonality and spatial distribution of rainfall in the state of Goiás and Federal District. Revista Brasileira de Geografia Física. 2012;**06**:1275-1291

[9] Swarowsky A, Heldwein AB, Buriol GA, D'avila RF, Estefanel V. Homogeneity and descriptive statistics of monthly and annual rainfall for Santa Maria, RS. Revista Brasileira de Recursos Hídricos. 2006;**11**:89-97. DOI: 10.21168/rbrh.v11n4.p89-97

[10] Searcy JK, Hardison CH. Double-mass curves. Manual of Hydrology: Part 1. General surface-water techniques. Washington; 1960. pp. 66

[11] Magalhães Filho FJC, Ayres FM, Sobrinho TA. Integrating GIS and USLE for soil loss mapping in environmental protection area. Revista Agrarian. 2014;**7**:552-559

[12] Chaves HML, Braga B Jr, Domingues AF, Santos DG. Quantificação dos benefícios ambientais e compensações financeiras do "Programa do Produtor de Água" (ANA): I.Teoria. Revista Brasileira de Recursos Hídricos. 2014;**9**:5-14

[13] Kinnell PIA. Event soil loss, runoff and the universal soil loss equation family of models: A review. Journal of Hydrology. 2010;**385**:384-397

[14] Wischmeier WH, Smith DD. Predicting Rainfall Erosion Losses: A Guide to Conservation Planning. U.S. Department of Agriculture, Handbook; 1978. 537

[15] Depiné H, Castro NZR, Pinheiro A, Pedrollo OC. Filling hourly data gaps in rainfall time series using artificial neural networks. Revista Brasileira de Recursos Hidrícos. 2014;**19**:51-63. DOI: 10.21168/rbrh.v19n1.p51-63

[16] Bertoni JC, Tucci CEM. Precipitação. In: CEM T, editor. Hidrologia: Ciência e Aplicação. 3th ed. Porto Alegre: UFRGS; 2002. p. 943

[17] Wissmann JA, Tampelini LG, Feil AC, Sampaio SC, Suszek M. Ferramenta computacional para análise de consistência de dados pluviométricos. Revista Varia Scientia. 2006;**06**:99-106

[18] Sanches FO, Balen DS, Silva RV, Rosa KK, Radunz AL. Rains in Rio Grande do Sul: A study of intense rainfall accumulated in the upper Uruguay Gaúcho. Revista Brasileira de Climatologia. 2014;**15**:143-162. DOI: 10.5380/abclima.v15i0.38074

[19] Chechi L, Sanches FO. Using the rainfall anomaly index (RAI) in evaluating the El Niño southern oscillation (ENSO) in the upper Uruguay gaucho from 1957 to 2012. Revista Brasileira de Geografia Física. 2013;**06**:1586-1597

[20] Lombardi Neto F, Moldenhauer WC. Rainfall erosivity–Its distribution and relationship with soil loss at Campinas, state of São Paulo, Brazil. Bragantia. 1992;**51**:189-196

[21] Oliveira PTS, Wendland E, Nearing MA. Rainfall erosivity in Brazil: A review. Catena. 2012;**100**:139-147. DOI: 10.1016/j.catena.2012.08.006

[22] Carvalho NO. Hidrossedimentologia Prática. 2nd ed. Rio de Janeiro: Interciência; 2008. p. 602

[23] Foster GR, McCool DK, Renard KG, Moldenhauer WC. Conservation of the universal soil loss equation to SI units. Journal of Soils and Water Conservation. 1981;**36**:355-359

[24] Trindade ALF, Oliveira PTS, Anache JAA, Wendland E. Spatial variability of rainfall erosivity in Brazil. Pesquisa Agropecuária Brasileira. 2016;**51**:1918-1928

[25] Kaschuk G, Alberton O, Hungria M. Quantifying effects of different agricultural land uses on soil microbial biomass and activity in Brazilian biomes: Inferences to improve soil quality. Plant and Soil. 2011;**338**:467-481. DOI: 10.1007/s11104-010-0559-z

[26] Ashouri H, Hsu KL, Sorooshian S, Braithwaite DK. PERSIANN-CDR: Daily precipitation climate data record from multisatellite observations for hydrological and climate studies. Bulletin of the American Meteorological Society. 2015;**96**:69-83. DOI: 10.1175/BAMS-D-13-00068.1

[27] Yong B, Liu D. Global view of real-time Trmm multisatellite precipitation analysis: Implications for its successor global precipitation measurement mission. Bulletin of the American Meteorological Society. 2015;**96**:283-296. DOI: 10.1175/BAMS-D-14-00017.1

[28] Yi S, Song C, Wang Q, Wang L, Heki K, Sun W. The potential of GRACE gravimetry to detect the heavy rainfall-induced impoundment of a small reservoir in the upper Yellow River. Water Resources Research. 2017;**53**:6562-6578. DOI: 10.1002/2017WR020793